EXPLOR

MATHEMATICS

With Maple

Kenneth H. Rosen
AT&T Bell Laboratories

John S. Devitt
Waterloo Maple, Inc.

Troy Vasiga
University of Waterloo

James McCarron
Waterloo Maple, Inc.

Eithne Murray
INRIA, Rocquencourt

Ed Rosko
AT&T Bell Laborator

The McGraw-Hill Com

New York   St. Louis   San Francisco   Auckland   I
Lisbon   London   Madrid   Mexico City   Milan   Mon
San Juan   Singapore   Sydney

McGraw-Hill

A Division of The McGraw-Hill Companies

Exploring Discrete Mathematics with Maple

2 3 4 5 6 7 8 9 0 FGR FGR 9 0 2 1 0 9 8 7 6

ISBN 0-07-054128-0

The editor was Maggie Rogers.
The production supervisor was Louis Swaim.
Fairfield Graphics was printer and binder.

Contents

Preface/Acknowledgments   **vii**

Introduction   **1**

Structure of This Volume 1

Interactive Maple 2

    An Example... 3

A First Encounter with Maple.. 5

    An Interest Rate Problem 6

Programming Preliminaries 9

    Getting Help.. 9

Basic Programming Constructions.. 11

    Iteration 11

    Structuring.. 14

    Branching 15

    Premature Loop Exit 17

Saving and Reading 17

Executing System Programs within Maple.. 18

Packages 18

Maple Versions 21

On-Line Material 21

    Using Anonymous FTP.. 22

Maple Worksheets 22

Communicating with the Authors.. 22

Exercises/Problems 23

1   Logic, Sets and Functions   **25**

    1.1   Logic 25

             Bit Operations 27

             Bit Strings 27

             A Maple Programming Example 28

             Loops and Truth Tables 29

             Using Maple to Check Logical Arguments............................... 32

    1.2   Quantifiers and Propositions 33

    1.3   Sets.. 36

             Set Operations 38

    1.4   Functions and Maple.. 41

             Tables... 41

             Functional Composition 46

    1.5   Growth of Functions 48

    1.6   Computations and Explorations 49

    1.7   Exercises/Projects 51

2   The Fundamentals   **53**

    2.1   Implementing Algorithms in Maple... 53

             Procedure Execution 54

             Local and Global Variables... 55

2.2 Measuring Time Complexity 59
2.3 Number Theory 62
Basic Number Theory 62
Greatest Common Divisors and Least Common Multiples 63
Chinese Remainder Theorem 66
Factoring Integers 67
Primality Testing 68
The Euler ϕ-Function 70
2.4 Applications of Number Theory 71
Hash Functions 71
Linear Congruential Pseudorandom Number Generators 75
Classical Cryptography 77
2.5 RSA Cryptography 79
Generating Large Primes 82
2.6 Base Expansions 83
2.7 Matrices 85
Zero-One Matrices 88
2.8 Computations and Explorations 91
2.9 Exercises/Projects 95

3 Mathematical Reasoning   99
3.1 Methods of Proof 99
3.2 Mathematical Induction 104
3.3 Recursive and Iterative Definitions 108
3.4 Computations and Explorations 111
3.5 Exercises/Projects 115

4 Counting   117
4.1 Relevant Maple Functions 118
4.2 More Combinatorial Functions 120
Binomial Coefficients 120
Multinomial Coefficients 123
Stirling Numbers 126
4.3 Permutations 128
Partitions of Integers 130
4.4 Discrete Probability 132
4.5 Permutations 133
4.6 Computations and Explorations 136
4.7 Exercises/Projects 144

5 Counting   147
5.1 Recurrence Relations 147
Towers of Hanoi Problem 148
5.2 Solving Recurrences with Maple 150
Inhomogeneous Recurrence Relations 157
Maple's Recurrence Solver 159

	Divide and Conquer Relations	163
5.3	Inclusion - Exclusion	164
5.4	Generating Functions	169
5.5	Computations and Explorations	173
5.6	Exercises/Projects	180

6   Relations   **183**
6.1	An Introduction to Relations in Maple..................................	183
6.2	Determining Properties of Relations using Maple	185
6.3	*n*-ary Relations in Maple	187
6.4	Representing Relations as Digraphs and Zero-One Matrices..............	190
	Representing Relations Using Directed Graphs	190
	Representing Relations Using Zero-One Matrices	192
6.5	Computing Closures of Relations	195
	Reflexive Closure..	195
	Symmetric Closure...	196
	Transitive Closure	196
6.6	Equivalence Relations...	199
6.7	Partial Ordering and Minimal Elements................................	201
6.8	Lattices..	205
6.9	Covering Relations...	206
6.10	Hasse Diagrams..	208
6.11	Computations and Explorations	212
6.12	Exercises/Projects	215

7   Graphs   **217**
7.1	Getting Started with Graphs...	217
	Simple Graphs...	217
	Visualizing Graphs in Maple...	219
	Directed Graphs	225
7.2	Simple Computations on Graphs..	229
	Degrees of Vertices	230
7.3	Constructing Special Graphs..	232
	Bipartite Graphs	237
	Subgraphs, Unions and Complements	240
7.4	Representing Graphs, and Graph Isomorphism	246
	Representing Graphs	246
	Graphs Isomorphism...	248
7.5	Connectivity	251
7.6	Euler and Hamilton Paths..	255
	Euler Circuits	255
7.7	Shortest Path Problems	258
7.8	Planar Graphs and Graph Coloring	261
	Planar Graphs	261
	Graph Colorings..	261
7.9	Flows...	263

CONTENTS

7.10   Computations and Explorations 265

7.11   Exercises/Projects 271

8   Trees   273

8.1   Introduction to Trees 273

      Rooted Trees 276

8.2   Application of Trees 280

      Binary Insertion 281

      Huffman Coding 286

8.3   Tree Traversal 288

      Infix, Prefix and Postfix Notation 292

8.4   Trees and Sorting 296

      Bubble Sort 297

      Merge Sort 298

8.5   Spanning Trees 300

      Backtracking 303

8.6   Minimum Spanning Trees 309

8.7   Computations and Explorations 314

8.8   Additional Exercises 320

9   Boolean Algebra   323

9.1   Boolean Functions 323

      A Boolean Evaluator 326

      Representing Boolean Functions 327

      Verifying Boolean Identities 327

      Duality 328

      Disjunctive Normal Form 329

9.2   Representing Boolean Functions 330

9.3   Minimization of Boolean Expressions and Circuits 333

9.4   Don't Care Conditions 336

9.5   Computations and Explorations 339

9.6   Exercises/Projects 345

10   Modeling Computation   349

10.1   Introduction 349

10.2   Stacks 350

10.3   Finite-State Machines with Output 353

10.4   Finite-State Machines with No Output 356

10.5   Deterministic Finite-State Machine Simulation 361

10.6   Nondeterministic Finite Automata 363

10.7   DFA to NFA 370

10.8   Converting Regular Expressions to/from Finite Automata 374

10.9   Turing Machines 378

10.10   Computations and Explorations 384

10.11   Exercises/Projects 387

Index

Preface

This book is a supplement to Ken Rosen's text *Discrete Mathematics and its Applications, Third Edition*, published by McGraw-Hill. It is unique as an ancillary to a discrete mathematics text in that its entire focus is on the computational aspects of the subject. This focus has allowed us to cover extensively and comprehensively how computations in the different areas of discrete mathematics can be performed, as well as how results of these computations can be used in explorations. This book provides a new perspective and a set of tools for exploring concepts in discrete mathematics, complementing the traditional aspects of an introductory course. We hope the users of this book will enjoy working with it as much as the authors have enjoyed putting this book together.

This book was written by a team of people, including Stan Devitt, one of the principle authors of the Maple system and Eithne Murray who has developed code for certain Maple packages. Two other authors, Troy Vasiga, and James McCarron have mastered discrete mathematics and Maple through their studies at the University of Waterloo, a key center of discrete mathematics research and the birthplace of Waterloo Maple Inc.

To effectively use this book, a student should be taking, or have taken, a course in discrete mathematics. For maximum effectiveness, the text used should be Ken Rosen's *Discrete Mathematics and its Applications*, although this volume will be useful even if this is not the case. We assume that the student has access to Maple, Release 3 or later. We have included material based on Maple shareware and on Release 4 with explicit indication of where this is done. (Where to obtain Maple shareware is described in the Introduction.) We do not assume that the student has previously used Maple. In fact, working through this book can teach students Maple while they are learning discrete mathematics. Of course, the level of sophistication of students with respect to programming will determine their ability to write their own Maple routines. We make peripheral use of calculus in this book. Although all places where calculus is used can be omitted, students who have studied calculus will find this material of interest.

This volume contains a great deal of Maple code, much based on existing Maple functions. But substantial extensions to Maple can be found throughout the book; new Maple routines have been added in key places, extending the capabilities of what is currently part of Maple. An excellent example is new Maple code for displaying trees, providing functionality not currently part of the network package of Maple. All the Maple code in this book is available over the Internet; see the Introduction for details.

This volume contains an Introduction and ten chapters. The Introduction describes the philosophy and contents of the chapters and provides an introduction to the use of Maple, both for computation and for programming. This chapter is especially important to students who have not used Maple before. (More material on programming with Maple is found throughout the text, especially in Chapters 1 and 2.) Chapters 1 to 10 correspond to the respective chapters of *Discrete Mathematics and its Applications*. Each chapter contains a discussion of how to use Maple to carry out computation on the subject of that chapter. Each chapter also contains a discussion of the of the Computations and Explorations found at the end of the corresponding chapter of *Discrete Mathematics and its Applications*, along with a set of exercises and projects designed for further work.

Users of this book are encouraged to provide feedback, either via the postal service or the Internet. We expect that students and faculty members using this book will develop material that they want to share with others. Consult the Introduction for details about how to download Maple software associated with this book and for information about how to upload your own Maple code and worksheets.

Acknowledgments

Thanks go to the staff of the College Division of McGraw-Hill for providing us with the flexibility and support to put together something new and innovative. In particular, thanks go to Jack Shira, Senior Sponsoring Editor and Maggie Rogers, Senior Associate Editor, for their strong interest, enthusiasm, and frequent inquiries into the status of this volume, and to Denise Schanck, Publisher, for her overall support. Thanks also goes to the production department of McGraw-Hill for their able work.

We would also like to express thanks to the staff of Waterloo Maple Inc. for their support of this project. In particular, we would like to thank Benton Leong and Ha Quang Le for their suggestions. Furthermore, we offer our appreciation to Charlie Colbourn of the University of Waterloo for helping bring this working team together as well as for his contributions as on of the authors of Maple's networks package which is heavily used in parts of this book.

As always, one of the authors, Ken Rosen, would like to thank his management at AT&T Bell Laboratories, including Steve Nurenberg, Ashok Kuthyar, Hrair Aldermishian, and Jim Day, for providing the environment and the resources that have made this book possible. Another author, Troy Vasiga, would like to thank his wife for her encouragement and support during the preparation of this book.

Introduction

An introduction to discrete mathematics is usually taught in a traditional way. Concepts, applications, and problem solving techniques are presented with worked examples provided to illustrate key points. The questions included in the textbook exercise sets to reinforce these key points are designed to be solved without a large amount of computation. But today, with modern mathematical computation software, such as Maple, complicated computations can be carried out easily and quickly. Having such computational tools provides a new dimension to a course in discrete mathematics course, namely an *enquiring* and *experimental* approach to learning. This book is designed to bridge the traditional approach to learning discrete mathematics and this new enquiring and experimental approach.

Using computational software, students can experiment directly with many objects important in discrete mathematics. These include sets, large integers, combinatorial objects, graphs, and trees. Furthermore, by using interactive computational software to do this, students can explore these examples more thoroughly thereby fostering a deeper understanding of concepts, applications, and problem solving techniques.

This supplement has two main goals. The first is to help students learn how to carry out computations in discrete mathematics using Maple, a widely used software package for interactive mathematical computation. The second is to guide students through the process of mathematical discovery through the use of computational tools. This exploration is based on the use of Maple.

Structure of This Volume

This supplement begins with a brief introduction to Maple, its capabilities and its use. Our goal is to provide students new to Maple with the requisite background. The material in this introduction explains the philosophy behind working with Maple, how to use Maple to carry out computations, and the basic structure of Maple. Maple is more than just a computational engine; it also is a programming language. Consequently, this introduction continues by explaining the basic constructs for programming with Maple. For those new to Maple, this material is important for understanding Maple procedures and their use.

Besides the introduction, the main body of this book contains ten chapters. Each chapter is based on a chapter of *Discrete Mathematics and its Applications*, third edition, by Kenneth H. Rosen, published by McGraw Hill (henceforth referred to as the *text*). A chapter begins with comprehensive coverage explaining how Maple can be used to explore the topics of the corresponding chapter of the text. This coverage includes a discussion of relevant Maple commands, as well as many new procedures, written expressly for this book. Many worked examples illustrating how to use Maple to explore topics in that chapter are provided. Additionally, a discussion of many of the Computations and Explorations in the corresponding chapter of the text is provided. Often, these exercises ask students to carry out a series of computations to explore a concept or study a problem. Here, we provide guidance, partial solutions, and sometimes complete solutions to these exercises. Finally, each chapter concludes with a set of additional questions for the students to explore. Some of these are straightforward computational exercises, while others are more accurately described as projects requiring substantial additional work, including programming. Consequently, this set of exercises is labeled Exercises/Projects.

The backmatter of this book includes an extensive index. Programming examples appear throughout the book, but even so, some supplementary programs were developed to facilitate the exposition and discussion. These supplementary programs are also listed.

This volume has been designed to help students achieve the main goals of a course in discrete mathematics. These goals, as described in the preface of the text, are the mastery of mathematical reasoning, combinatorial analysis, discrete structures, applications and modeling, and algorithmic thinking. This supplement demonstrates how to use the interactive computational environment of Maple to enhance and accelerate the achievement of these goals.

Interactive Maple

Exploring mathematics with Maple is like exploring a mathematical topic with an expert assistant at your side. As you investigate a topic you should always be asking questions. In many cases the answer to your question lies in an experiment. Maple, your highly trained mathematical assistant, can often carry out these directed experiments quickly and accurately. Often this requires only a few simple directives (commands).

By hand, the magnitude and quantity of work required to investigate even one reasonable test case may be prohibitive. By delegating the detail to your mathematical assistant your efforts are much more focussed on choosing the right the mathematical questions and on interpreting their results than in a traditional computational environment.

The reasons an interactive system such as Maple supports such rapid investigation include the fact that:

1. The types of objects you are investigating already exist as part of the basic infrastructure provided by the system. This includes sets (ordered and unordered), variables, polynomials, graphs, arbitrarily large integers and rational numbers, and most importantly, support for exact computations.

2. Tools for manipulating those objects already exist, or can be created easily by essentially mimicking the interactive solution to a particular problem.

Current day calculators are appropriate for simple numerical investigations, but they do not allow you to effectively investigate more complicated mathematical structures or to quickly prototype multistep methods for manipulating those complicated structures.

On the other hand, more traditional programming languages, the kind used to build Maple, almost all of your effort goes into just building an appropriate environment (potentially years of work).

An Example

The use of Maple is merely a means to the end of achieving the goals of a course in discrete mathematics. Yet, with any tool, to use it effectively you must have some basic understanding of the tool and its capabilities. In this section we introduce Maple by working through a sample interactive session.

A new Maple session begins by starting the Maple program. It presents you with a blank *Maple worksheet*, much as any word processor would begin by presenting you with a blank page. You can continue an old session by "opening" an old document, or by specifying the name of such a document at the time you start Maple.

The Maple session then proceeds by entering Maple comments, and studying Maple's responses (which appear in the document) and inserting comments. The comments are entered as paragraphs, much as in any word processor. The document is called a Maple worksheet.

For the most part, a Maple worksheet behaves much like any word processor document. You create descriptions or commands by typing, backspacing, and generally editing the contents of the document using all the standard operations of a word processor.

It differs from a word processor document in that selected paragraphs (or lines of input) can be designated as being *command input regions.* These command input regions are somewhat restricted in that they must contain only valid Maple commands.

However, if you press ENTER anywhere on a line containing a complete Maple command the command is handed off to the Maple computational engine, the requested computations are carried out, and then inserted into the document in the paragraph immediately following the command.

Thus, it is just like editing a document with an eager mathematical assistant at your side, waiting for you to delegate various mathematical tasks.

Typically, upon starting Maple, the cursor is placed on a new line beginning with a special prompt character > . This "prompt" indicates that you are in a command region. You proceed by typing your command and pressing the Enter key.

In the sample session which follows, the Maple commands are shown in a distinct font on lines beginning with the special prompt character > . Typically, they consist of mathematical formulae together with a function indicating what action or transformation is to be performed, and ending in a semi-colon.

Complete commands end in semi-colons (;). This is so that more than one maybe grouped on one line. In the absence of a terminating semi-colon Maple normally expects the command to be completed on the next line.

To execute a command, make sure that the insertion point is some where on the line containing the command, and press Enter to compute (or re-compute) the result and display the answer.

Symbolic algebra systems are impressive in terms of the range and type of computations they can do.

Three typical commands[1] and their results are as follows.

```
>   Sum( (i+5)^3,i=1..n);
```

---

[1]The semi-colons are essential parts of the commands (serving to indicate completion). Also, In the last two commands, the double quote " is used to indicate the "value of the last computed result".

$$\sum_{i=1}^{n} (i+5)^3$$

```
>   expand(");
```

$$\sum_{i=1}^{n} (i^3 + 15\,i^2 + 75\,i + 125)$$

```
>   value(");
```

$$\frac{1}{4}\,(n+1)^4 + \frac{9}{2}\,(n+1)^3 + \frac{121}{4}\,(n+1)^2 + 90\,n - 35$$

The result produced by each command is inserted immediately following the actual command.

A First Encounter with Maple

As already indicated, working with Maple is like working with an expert mathematical assistant. This requires a subtle change in the way you think about a problem. Instead of focusing on "how do I do XYZ?" your primary role becomes that of deciding "what" needs to be done next.

Much of discrete mathematics is about understanding the relationship between actual objects or sets of objects and mathematical models that are set up to capture some property of these objects. Understanding these relationships often requires that you view either the objects or the associated mathematical model in different ways.

Maple allows you to manipulate the mathematical models almost casually. For example, the polynomial $(x + (x + z)y)^3$ can be entered into Maple as the command

```
>   (x + (x+z)*y )^3;
```

$$(x + (x + z)\,y)^3$$

The result of executing this "command" is displayed immediately after the command. In this case Maple simply echoes the polynomial as no special computations were requested.

The power of having a mathematical assistant now becomes apparent because a wide range of standard operations become immediately available. For example, you can expand, differentiate, integrate simply by deciding that this is what is needed. Maple uses (") to refer to the value of the previous computation. Thus a full expansion of the previous polynomial takes place in response to the command

```
>  expand(");
```
$$x^3 + 3\,y\,x^3 + 3\,x^2\,y\,z + 3\,y^2\,x^3 + 6\,y^2\,x^2\,z + 3\,x\,y^2\,z^2 + y^3\,x^3 + 3\,y^3\,x^2\,z$$
$$+ 3\,y^3\,x\,z^2 + y^3\,z^3$$

Perhaps you wish to view it as if it were actually a polynomial in x with the y's and z's placed in the coefficients. To collect the previous polynomial as a polynomial in x, use the command

```
>  collect( " , x );
```
$$(y^3 + 3\,y^2 + 1 + 3\,y)\,x^3 + (3\,y\,z + 3\,y^3\,z + 6\,y^2\,z)\,x^2 + (3\,y^3\,z^2 + 3\,y^2\,z^2)\,x$$
$$+ y^3\,z^3$$

To return to a factored form, simply request that the previous result be factored.

```
>  factor(");
```
$$(x + y\,x + y\,z)^3$$

The flexibility to move quickly between different representations of the same object will prove to be extremely useful when looking for a solution to a problem.

Armed with this kind of expert mathematical assistant, experiments (even complicated ones) can be run quickly so that you are freer to explore ideas and models.

A second very important benefit is that the particular computations that you choose to perform (even lengthy ones) are performed accurately. Thus, the feedback you do get from your experiments is much more likely to be feedback on the model you had chosen rather than nonsense arising because of a simple arithmetic error.

Finally, the sheer computational power of such an assistant allows you to run much more extensive experiments. This can be important when trying to establish or identify a relationship between a mathematical model and a collection of discrete objects.

An Interest Rate Problem

To see how working with a computational assistant like Maple can help with the problem solving process let us build a model for computing compound interest at (say) 10% interest per annum, compounded annually.

In almost every investigation it is helpful to start with a particular example. Thus we consider a case where the initial balance is given by:

```
>  p(0) = 1000;
```
$$p(0) = 1000$$

We will want to refer to this equation again so we give it a name, say eq0 by using the name assignment operator := . (Remember that (") refers to the previously evaluated expression!)

```
>  eq0 := " ;
```
$$eq0 := p(0) = 1000$$

The interest rate can be given similarly by the (named) equation eq1 as in

```
>  eq1 :=   ( i = .1 );
```
$$eq1 := i = .1$$

The total deposit p(1) at the end of the first interest period is related to p(0) by the equation

```
>  p(1) =   p(0) + i * p(0);
```
$$p(1) = p(0) + i\,p(0)$$

To compute a numerical value for p(1), we simply use the equation eq0 and eq1 to substitute in a specific values for i and $p(0)$. This is easily done via the subs command as in

```
>  subs( {eq0, eq1} , " );
```
$$p(1) = 1100.0$$

We now have a specific solution for one interest period. Also, we have invested some effort in setting up the problem using the correct notation for the mathematical assistant. How does this help us?

1. We can now easily re-compute specific examples for different values of $p(0)$ and i. Such recomputations are often used to gain more insight or to test the model for extreme cases. For example if the interest rate is 0, we would expect p(1) and p(0) to be equal. (This kind of test helps us to verify that our model makes sense.) To verify this here, go back to the command line defining eq1 and by selecting and typing, change the value of i to 0. Then simply hit the ENTER key several times until a new value for p(1) is computed.

   Note that without further typing or data entry, each time you hit ENTER, the current command is executed and the cursor advances to the next command. This allows you to run through the steps of your emerging model quickly.

2. By simply deleting the Maple results from the worksheet we are left with a high level summary of how to solve the problem. (Use the FORMAT menu to do this now.)

The assistant can take us even further with this problem, without bogging down in details. Given any particular period, say the nth period, the value of the principal is then $p(n)$. The value of the principal at the end of the next interest period is $p(n+1)$. These two values are related in essentially the same way as equation eq1. In Maple notation the equation is

```
>  p(n+1) = p(n) + i*p(n);
```

$$p(n + 1) = p(n) + i\,p(n)$$

This kind of equation is often referred to as a *recurrence equation*. In Chapter 5 we will discuss a general approach to solving recurrence equations. However, our expert assistant Maple is already able to carry this out. Thus, if we ignore the details of finding the solution, the `rsolve` command can be used to generate the general solution to this problem. The result is:

```
>  eqn :=  p(n) = rsolve( " , p(n) );
```

$$eqn := p(n) = p(0)\,(1 + i)^n$$

This can be used to create a function for computing the principle after n steps. The formula for $p(n)$ is:

```
>  subs( eqn , p(n) );
```

$$p(0)\,(1 + i)^n$$

A function for computing this as a function of n and i is:

```
>  f :=   unapply ( " , n , i );
```

$$f := (n, i) \rightarrow p(0)\,(1 + i)^n$$

Thus, the principal after 10 interest periods at 18% is:

```
>  f( 10 , .18 );
```

$$5.233835554\,p(0)$$

At $p0 = 1000$, this is:

```
>  subs( p(0) = 1000 , " );
```

$$5233.835554$$

Programming Preliminaries

This section is intended for those readers who may have little or no previous exposure to programming. We'll endeavor to provide you with enough information to get you started, so that you can work productively with Maple. Limited space prevents a really thorough treatment of these topics, however. For further information, you are encouraged to consult the Maplemanuals, which will provide you with detailed explanations of all of Maples programming facilities, and further examples of their use.

Getting Help

The first thing that you need to know is how to use the `help` facility of Maple. Throughout this book a variety of Maplelibrary procedures are used and, while most are explained in the context of their immediate use, there is often much more information available online than would be appropriate to cover in this volume.

Suppose that you would like to know more about a Maple command called `foo`. You can access the online help for `foo` within your Maple session by simply typing a question mark followed by the name `foo`.

```
>   ?foo
```

```
%% Help error, ...
```

No help is displayed in this example because, currently, there is no Maple command called "foo". There is a Maple command called `proc`, however, and you will use it many times in the course of working through this book. To read the online help for `proc` you type

```
>   ?proc
```

The result is a help page as in:

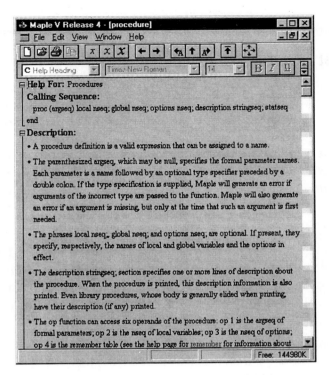

The exact manner in which the help is displayed for you depends on the type of interface you are using. It may also depend on the version of Maple that you are using. (One of the features of Release 4 is hyperlinked online documentation.) Each help page for Maple commands has several sections:

HELP FOR states what is being described

CALLING SEQUENCE how to use the function, with arguments

DESCRIPTION description, in point form, of the purpose and functionality provided by the function

EXAMPLES (not shown above) examples of usage

SEE ALSO (not shown above) cross references to other online help pages

(Not all sections will be present in every help page.) Every Maple command is documented in the online help. You can use it to find all the available information that you need to use the command. In addition to library procedures, various keywords and programming constructs, such as `for` and `if` are documented in the online help system.

Basic Programming Constructions

All programming languages provide a few basic means for the construction of algorithms. Traditional programming languages such as C or Lisp provide various mechanisms for controlling the flow of logic in a programmatic description of an algorithm, and Maple provides easy to use mechanisms for flow control as well. While the syntax varies from one language to the next, there are certain elements that are present in most languages in one form or another. This section will describe the following means for the specification of algorithms in Maple.

1. iteration

2. structuring

3. branching

Iteration

Mapleprovides several means of iteration. Iteration is a mechanism for doing some task repeatedly. In most programming languages, Maple included, this is accomplished in some sort of looping structure.

Iteration Over an Arithmetic Sequence

The most basic type of iteration of the *for loop*. Mapleactually provides two kinds of for loops. The first allows you to iterate over an arithmetic progression.

```
>   for i from 3 to 9 by 2 do
>     i * i;
>   od;
```

$$9$$
$$25$$
$$49$$
$$81$$

Reading these three lines aloud will give you a pretty good idea of what they accomplish. A variable i is begin used as a counter. At the start of the loop, i is given the value 3. Each pass through the loop increments the value of i by 2, until i has the value 9. Each iteration of the loop caused the execution of the line

```
>   i * i;
```

There is no limit on the amount of calculation that is done inside the loop. The contents of the loop can span as many lines as you like. For this reason, the final od; is required to tell Maple where the inside of the loop ends.

Iteration Over Members of a List

The second form of `for` loop allow you to iterate over the members of a list. To perform the same task as the loop above, but using a list instead, we can type

```
>  mylist := [3,5,7,9]:
>  for i in mylist do
>    i * i;
>  od;
```

$$9$$
$$25$$
$$49$$
$$81$$

The assignment statement `mylist := [3,5,7,9]:` was ended with a colon to prevent the list from being displayed.

Alternatively, a set can be used

```
>  myset := {3,5,7,9}:
>  for i in myset do
>    i * i;
>  od;
```

$$9$$
$$25$$
$$49$$
$$81$$

The advantage of being able to iterate over a list or a set is that it is much more general. For instance, the numbers $1, 5, 98, 345$ do not form an arithmetic progression, yet we can still iterate over them if we construct a list with these numbers as members.

```
>  mylist := [1,5,98,345];
```

$$mylist := [1, 5, 98, 345]$$

```
>  for i in mylist do
>    i * i;
>  od;
```

$$1$$
$$25$$
$$9604$$
$$119025$$

This also allows you to iterate over collections of other than numbers. In fact, you can iterate over any collection of Maple objects.

More General Iteration

Somewhat more general than a for loop is a *while loop*. This is another method for iteration provided by the Maple interpreter. Here is an example.

```
>   i := 3:
>   while i <= 9 do
>     print( i * i);
>     i := i + 2;
>   od:
```

$$9$$
$$25$$
$$49$$
$$81$$

Note that loops, like assignment statements, can be ended with a colon. For loops, this suppresses the automatic printing of all the statements inside the loop. Then, the `print` command can be used to display selected results.

Just as with the `for` loops described previously, this loop calculates the square of the odd integers between 3 and 9, inclusive. Notice that extra steps are required here that were implicit in each of the `for` loops above. First, the value of the variable i needs to be initialized outside of the loop. Second, an explicit step within the loop is needed to increment the loop variable i. But, this type of loop provides greater generality. With a `while` loop, we can loop over an arbitrary condition. For example

```
>   i := 1:
>   while i^7 < 150 do
>     print(i);
>     i := i + i^2;
>   od:
```

$$1$$
$$2$$

Note that in this example, only the value of i is shown. Furthermore, it is not necessary that the loop variable be a numeric object. Loop variables can be any type of Maple object. Later in this book we'll see examples of loops in which the loop variable is of a non-numeric type, such as a set or an edge of a graph.

Maplehas the rare feature that you can *combine* the `for` and `while` loop semantics into a single loop construction. Here is an example of how this can be done.

```
>   for i from 3 to 44 while i^2 < 50 do
>     i, i^2;
>   od;
```

$$3, 9$$

$$4, 16$$
$$5, 25$$
$$6, 36$$
$$7, 49$$

Structuring

Perhaps the most important programming construction in Maple — and in most programming languages — is the *procedure*. A Maple procedure is very much like a function in mathematics. It is an object that is capable of receiving data as *input*, and producing data as *output*. Unlike a mathematical function, however, it may also perform some action as a side effect; that is, it may change its input, or some global data as a result of its execution.

A Maple procedure is created using the `proc` keyword. Normally, but not necessarily, procedures are assigned to a variable; this gives the procedure a name, so that it can be used later. Let's consider a simple example.

```
>   MySum := proc(a,b)
>     a + b;
>   end;
```

$$MySum := \mathbf{proc}(a, b)\, a + b \ \mathbf{end}$$

This creates a procedure — the part appearing between the `proc` and the `end` — and assigns it to the variable `MySum`. Note that the naming of a Maple procedure via the `:=` operator is identical the naming of any other Maple object. The variables `a` and `b` are *parameters* or *arguments* to the procedure `MySum`. The output of the procedure is the result of the last statement that is executed in the body of the procedure. Sometimes (see the next example) this will be indicated by a `RETURN` statement. (The *body* of the procedure is the sequence of statements that occur between the `proc` keyword with its parameter list and the `end` keyword.)

The ending semicolonis not an essential part of the procedure. It is Maple's way of terminating any statement. Colons will usually be used (suppressing display of the procedure definition) since we have no need to see procedure definition twice.

The primary purpose of procedures is to encapsulate algorithms. Because of their importance, procedures will be discussed in detail in Chapters 1 and 2. We introduce them here very briefly merely to give meaningful context to the next topic, *branching*.

Branching

Branching in Maple is achieved through the use of the if..then..else..fi structure. It is a mechanism that allows you to choose between one or more events based on conditions that can only be detected during the course of a computation. It also allows you to modify in certain ways the behavior of loops. In its simplest form, the "if statement" allows you to decide whether to take some action or not.

```
>  if isprime(13) then
>    print(13, 'is prime');
>  fi;
```

$$13, \textit{is prime}$$

Just as a loop body that is introduced with the keyword do must be terminated with the keyword od, an if statement has a body that must be terminated with the keyword fi, and is introduced by the keyword then. If you type in the first example above during an interactive Maple session, you will see that nothing happens until you have finished typing the line containing the fi keyword. The body of an if statement is executed only if the condition of the statement yields a true value, as it does in the first of our two examples. (In the two examples above, the *condition* is a test for integral primality that, in this case, uses the Maple library function isprime.) Should the condition produce a false value, the body of the if statement is simply ignored, as was the case with the second example above.

These examples are a little unrealistic because in a toplevel interactive Maple session you would test the primality of 13 simply by typing

```
>  isprime(13);
```

$$\textit{true}$$

An if statement will occur most commonly within the body of a procedure definition, and the condition will be formed using the parameters of the procedure.

```
>  TestPrime := proc(n)
>    if isprime(n) then
>      print(n, 'is a prime number');
>      RETURN(true);
>    fi;
>    RETURN(false);
>  end:
```

A message is *printed*, and a return valueindexreturn valueis provided. The print statement has no value. Rather, the return value is either **true** or false, depending on which of the RETURN statements is encountered first. Note how it is used.

```
> TestPrime(13);
```

$$13, \textit{is a prime number}$$
$$\textit{true}$$

```
> TestPrime(16);
```

$$\textit{false}$$

Often, you will want to take one action if a condition is true and another if a condition is false. The else keyword allows you to extend an if statement to contain a second alternative. Here is an example that extends the TestPrime procedure above.

```
> TestPrime := proc(n)
>    if isprime(n) then
>      print(n, 'is a prime number');
>    else
>      print(n, 'is not a prime number');
>    fi;
> end:
> TestPrime(16);
```

$$16, \textit{is not a prime number}$$

You can also extend the if statement to a multiway branching structure. To do this, you use the keyword elif and test against several conditions. Here is an example of the syntax.

```
> TestDivisor := proc(n)
>    if irem(n,2) = 0 then
>      print(n, 'is divisible by 2');
>    elif irem(n, 3) = 0 then
>      print(n, 'is divisible by 3');
>    elif irem(n, 4) = 0 then
>      print(n, 'is divisible by 4');
>    elif irem(n, 5) = 0 then
>      print(n, 'is divisible by 5');
>    fi;
>    RETURN(NULL);
> end:
```

Here, we are using the Maple library procedure irem, which computes the remainder left upon division of its first argument by its second. This procedure only prints messages. The special name NULL ensures that the procedure does not return any value.

Premature Loop Exit

Sometimes it is necessary to terminate the body of a loop prematurely. This may occur because an error has occurred, or because the logic of a particular problem dictates that it must. For this, the break keyword can be used to transfer control out of a loop. For instance, here is one way to find the first prime number that occurs after a given number.

```
>   PrimeAfter := proc(n)
>     local i;
>     i := 1;
>     while i < 1000 do
>       if isprime(n + i) then
>         break;
>       else
>         i := i + 1;
>       fi;
>     od;
>     RETURN(n + i);
>   end:
```

This example is rather artificial because break is rarely used except in fairly complex algorithms with several nested levels of loops, but it serves to illustrate the syntax.

Saving and Reading

When you are developing long or complex procedures, or even a library of procedures, it is usually convenient to store the procedure definitions in a file on your computers disk. This makes it possible to reload the definitions the next time you run Maple. To save your entire Maple session to a disk file, choose a name for the file, such as mysession and type the command i¿ save 'mysession'; The next time you start a Maple session, you can restore the state of your previous session by loading the file using the command read. i¿ read 'mysession'; If you only wish to save the definition of one particular function, say PrimeAfter, you can type i¿ save 'PrimeAfter', 'mysession'; Strictly speaking, the back quotes are not necessary in these example, but it is good to form the habit of using them. They are necessary if you want to use a file name such as mysession.txt that contains non-alphanumeric characters other than the underscore. Note that it is not necessary to create Maple procedures within a Maple session. You can type Maple commands into a text file using an ordinary text editor such as vi or emacs. (Note, however, that you **cannot** use a word processor, unless it has the ability to save the file as a plain text file.) Then you can load those definitions into a Maple session using the read command.

Executing System Programs within Maple

Often, during a Maple session, you need to execute a system command on the computer, but do not wish to stop the Maple session to get back to the shell. Mapleprovides the `system` command, which allows you to execute system programs from within a running Maple session. To do this, you pass the command to execute as a string argument to the `system` command. For example, to list the files in a specific directory "project", you can type

```
> system('ls project'):
> chapter.tex
> over1.tex
> over2.tex
> packages.tex
```

It is also possible to use `system` within a procedure definition. Here, for example, is a simple, but handy, procedure that you can use to develop Maple source files.

```
> Fix := proc(filename::string)
>    system('vi '.filename); # edit the file
>    read(filename); # read it when finished
> end:
```

To use it to work on a file named `myfile.maple`, you can type

```
> Fix('myfile.maple');
```

This will call the `vi` editor on the file and, when you are finished making edits to the file, it will read the file into the Maple session, and then return you to the interactive session where you can continue testing or using its contents. Of course, you should replace `vi` with your favorite text editor.

System Architecture and Packages

This section briefly explains the global structure of the Maple system. It is intended to help you better understand how Maple works, and why some things work the way that they do.

Maple uses an innovative system architecture to achieve ambitious design goals. The Maple *kernel* is written in the C programming language, and implements the basic interpreter of the Maple programming language, the interface with the host computer's operating system, and certain time critical services.

However, nearly *all* of Maple's mathematical knowledge and power dwells in the massive Maple *library.* Consisting of thousands of lines of code, the Maple library is written in the Maple programming language itself. The advantages of this design include:

Portability As new computer systems emerge on the market, only the relatively small kernel needs to be ported. (*Porting* refers to the process by which a program is modified to run on a new system.)

Extensibility Using a system in which much functionality is loaded at runtime permits users to add their own extensions to the system. You can even *re*define existing Maple library routines (very carefully!) to extend or modify their behavior. This was done, for instance, in the production of this book, so that the `draw` command used to display graphs now includes new functionality.

Openness Users of Maple can gain greater understanding of the algorithms they are using by reading most of the library source code that implements those algorithms.

Because of its size, much of the Maple library is organized into subsets called *packages.* A package is a collection of Maple library routines that offer related services. Since these are not normally loaded when you start Maple, you must request the services localized in each package by telling Maple explicitly that you want to use them. For this purpose, Maple provides the `with` command; it is used to load a specific package.

For example, the `combinat` package, provides routines related to combinatorics. To use procedures from the `combinat` package, you would first load the package by typing:

```
> with(combinat);
```
[Chi, *bell*, binomial, *cartprod*, *character*, *choose*, *composition*, *conjpart*, *decodepart*, *encodepart*, *fibonacci*, *firstpart*, *graycode*, *inttovec*, *lastpart*, *multinomial*, *nextpart*, *numbcomb*, *numbcomp*, *numbpart*, *numbperm*, *partition*, *permute*, *powerset*, *prevpart*, *randcomb*, *randpart*, *randperm*, *stirling1*, *stirling2*, *subsets*, *vectoint*]

All the names of the newly loaded library routines are listed. To suppress this listing, use a colon (:) instead of a semicolon (;) as the statement terminator.

Depending upon which package you are loading, and upon which packages are already loaded, you may see one or more warnings that certain names have been redefined. This is *not* an indication that anything is wrong. The warnings merely serve as a reminder to the user that some procedure names have acquired new meaning as a result of having introduced a new package into the Maple session. This can occur because there may be two functions in different packages with the same name. Such a conflict exists, for example, between two very different functions called `fibonacci`, one in the `linalg` package, and one in the `combinat` package. You can still use both functions in the same Maple session, however, because each procedure has a "full name" consisting of the name of the package and the name of the procedure itself. For example, if you have loaded the `combinat` package, but you want to use the `fibonacci` routine found in the `linalg` package, you can type

```
> linalg[fibonacci](3);
```

$$\begin{bmatrix} 1 & 1 & 1 \\ 1 & 0 & 1 \\ 1 & 1 & 0 \end{bmatrix}$$

to compute the 3rd Fibonacci matrix. You can use the long form of the name even if you have not loaded the `linalg` package prior to using the `with` command.

The Maple library has many packages; not all of them will be prominent in this book, which concentrates on discrete mathematics. Some of the packages most relevant to this book include:

combinat: combinatorial functions

networks: routines for working with graphs and digraphs

linalg: linear algebra functions

logic: procedures for the symbolic manipulation of boolean expressions

numtheory: various number theoretic functions and procedures for working with integers.

plots: plotting routines

powseries: functions for manipulating formal power series

A new package names `combstruct`, discussed in Chapter 4, has been added to Release 4 of Maple. It is also available as a share package in Release 3.

You can see a list of all the available packages by typing `?packages` at the Maple prompt. You can read more about a particular package by using the help system described earlier.

Maple Versions

The procedures and examples in this book have been developed and tested under both Maple V Release 4 and Maple V Release 3. In some cases, minor syntactic changes are required in order to re-build a procedure under Release 3. The most notable example is in procedure definitions where a definition of the form

```
> f := proc(x :: integer ) ...
```

must be written as

```
> f := proc(x : integer ) ...
```

in Release 3.

On-Line Material

The source code for the examples and procedures developed in this book as well as information pertaining to any third party packages that are used in the various sections are available on-line courtesy of Waterloo Maple Inc.

The new procedures specific to this book are organized as a library of Maple packages. This special library, and the related examples and code are organized into files and directories directories found at

```
ftp.maplesoft.com
```

and available via *anonymous ftp*. The relevant files at the ftp site are found in the subdirectory

```
pub/maple/books/rosen
```

The `readme` file in that directory contains detailed instructions on how to access and use the online examples for both Release 3 and Release 4 of Maple. You are encouraged to contribute your own examples. Instructions on how make electronic contributions are mentioned in the `readme` file.

You can also access these files using your favorite Web browser. Simply enter one of the indicated *URL* addresses.

```
ftp://ftp.maplesoft.com
http://www.maplesoft.com
```

Browser commands can then be used to view or download copies of the files from the ftp site.

Using Anonymous FTP

To access these files via *anonymous ftp* specify the system name. On most UNIX systems this is accomplished by the command

```
% ftp ftp.maplesoft.com
```

Respond to the request for a user name with the name "anonymous" . Respond to the request for a password with your own email address, for example, "myuserid@myhost.edu".

Questions concerning this ftp site should be sent via electronic mail to doc@maplesoft.com.

Maple Worksheets

On many systems Maple Release 4 supports an elaborate windowing interface through documents called *Worksheets*. Worksheets provide many advanced editing and processing features such as hyptext linking, styles and outlining. Many of the on-line examples found at the ftp site for this book are presented in this format.

Details of the interface can be found by selecting the *Table of Contents* entry from the *help menu* and then clicking on the *Worksheet Interface* entry in the help document that is opened.

You can and should use worksheets to document your own solutions to problems. The resulting documents can be emailed without modification.

Communicating With The Authors

If you would like to communicate with the authors of this book, you can send your message to books_rosen@maplesoft.com. We encourage you to send your suggestions, comments, and material you might want to share with others to us at this address. We are interested in new procedures developed by students and instructors, as well as original worksheets. Contributions can be uploaded to the `incoming` directory at the ftp site or emailed as worksheets to the authors. Detailed instructions on how to make contributions are presented in the `readme` file.

Exercises/Problems

Exercise 1. Write the following `for` loop as a `while` loop.

```
>   for i from 1 to 10 do
>     sqrt(i);
>   od;
```

Exercise 2. Rewrite the `while` loop

```
>   i := 10;
>   while i^3 < 100 do
>     i;
>   od;
```

as a `for` loop.

1   The Foundations: Logic, Sets, and Functions

This chapter describes how to use Maple to study three topics from the foundations of discrete mathematics. These topics are logic, sets, and functions. In particular, we describe how Maple can be used in logic to carry out such tasks as building truth tables and checking logical arguments. We show to use Maple to work with sets, including how to carry out basic set operations and how to determine the number of elements of a set. We describe how to represent and work with functions in Maple. Our discussion of the topics in this chapter concludes with a discussion of the growth of functions.

1.1   Logic

The values of true and false (T and F in Table 1 on page 3 of the main text) are represented in Maple by the words `true` and `false`.

```
>  true, false;
```
$$true, false$$

Names can be used to represent propositions. If the truth value of p has not yet been determined, its value is just p. At any time, you can test this by entering the Maple expression consisting of only the name, as in

```
>  p;
```
$$p$$

A value can be assigned to a name by using the := operator.

```
>  p := true;
```
$$p := true$$

Subsequently, every reference to p returns the new value of p, as in

```
>  p;
```
$$true$$

The value of p can be removed by assigning p its own name as a value. This is done by the statement

```
>  p := 'p';
```
$$p := p$$

The quotes are required to *stop* p from evaluating.

The basic logical operations of *negation*, *Conjunction* (and), and *Disjunction* (or), are all supported. For example, we can write:

```
>  not p;
```
$$\textbf{not } p$$

```
>  p and q;
```
$$p \textbf{ and } q$$

```
>  p or q;
```
$$p \textbf{ or } q$$

None of these expressions evaluated to `true` or `false`. This is because evaluation can not happen until more information (the truth values of p and q) is provided. However, if we assign values to p and q and try again to evaluate these expressions we obtain truth values.

```
>  p := true:  q := false:
>  not p;
```
$$\textit{false}$$

```
>  p and q;
```
$$\textit{false}$$

```
>  p or q;
```
$$\textit{true}$$

Maple does not support operations such as an *exclusive or* directly, but it can be easily programmed. For example, a simple procedure that can be used to calculate the *exclusive or* of two propositions is defined as:

```
>  XOR := proc(a,b) ( a or b ) and not (a and b) end:
```

It is a simple matter to verify that this definition is correct. Simply try it on all possible combinations of arguments.

```
>  XOR( true , true );
```
$$\textit{false}$$

```
>  XOR( true , false );
```
$$\textit{true}$$

With the current values of p and q, we find that their exclusive or is true.

```
>  XOR( p , q );
```
$$\textit{true}$$

Bit Operations

We can choose to represent *true* by a 1 and *false* by a 0. This is often done in computing as it allows us to minimize the amount of computer memory required to represent such information.

Many computers use a 32 bit architecture. Each bit is a 0 or a 1. Each word contains 32 bits and typically represents a number.

Operators can be defined similar to **and** and **or** but which accept 1s and 0s instead of true and false. They are called bitwise operations. The bitwise *and* operator, AND, can be defined as

```
>   AND := proc( a , b )
>     if a = 1 and a = b then 1
>     else 0 fi;
>   end:
```

For example, the binary value of AND(0,1) is:

```
>   AND(0,1);
```

$$0$$

Bit Strings

Once defined, such an operation can easily be applied to two lists by using the bitwise operation on the pair of elements in position 1, the pair of elements in position 2, and so on. The overall effect somewhat resembles the closing of a zipper and in Maple can be accomplished by using the command zip. For example, given the lists

```
>   L1 := [1,0,1,1,1,0,0]: L2 := [1,1,1,0,1,0,1]:
```

we can compute a new list representing the result of performing the bitwise operations on the pairs of entries using the command

```
>   zip( AND , L1 , L2 );
```

$$[1, 0, 1, 0, 1, 0, 0]$$

Beware! This direct method only works as intended if the two lists initially had the same length. The zip command is used when you want to apply a function of two arguments to each pair formed from the members of two lists (or vectors) of the same length. In general, the call zip(f, u, v), where u and v are lists, returns the list [f(u[1], v[1]), f(u[2], v[2]), ... , f(u[length(u)], v[length(v)])]. It allows you to extend binary operations to lists and vectors by applying the (arbitrary) binary operation coordinatewise.

A Maple Programming Example

Using some of the other programming constructs in Maple we can rewrite AND to handle both bitwise and list based operations and also take into account the length of the lists.

We need to be able to compute the length of the lists using the nops command as in

```
>  nops(L1);
```
$$7$$

to take the maximum of two numbers, as in

```
>  max(2,3);
```
$$3$$

and to be able to form new lists. We form the elements of the new lists either by explicitly constructing the elements using the seq command or by using the op command to extract the elements of a list. The results are placed inside square brackets to form a new list.

```
>  L3 := [ seq( 0 , i=1..5) ]; L4 := [ op(L1) , op(L3) ];
```
$$L3 := [0, 0, 0, 0, 0]$$
$$L4 := [1, 0, 1, 1, 1, 0, 0, 0, 0, 0, 0, 0]$$

We can use this to extend the length of short lists by adding extra 0s.

In addition, we use an "if ... then" statement to take different actions depending on the truth value of various tests. The type statement in Maple can test objects to see if they are of a certain type. Simple examples of such tests are:

```
>  type(3,numeric);
```
$$true$$

```
>  type(L3,list(numeric));
```
$$true$$

```
>  type([L3,L4] , [list,list] );
```
$$true$$

A new version of the AND procedure is shown below.

```
>  AND := proc(a,b)
>    local i, n, newa, newb;
>    if type([a,b],[list,list]) then
>      n := max( nops(a),nops(b) );   # the longest list.
>      newa := [op(a) , seq(0,i=1..n-nops(a)) ];
>      newb := [op(b) , seq(0,i=1..n-nops(b)) ];
>      RETURN( zip(AND,newa,newb) )
>    fi;
>    if type( [a,b] , [numeric,numeric] ) then
>      if [a,b] = [1,1] then 1 else 0 fi
>    else
>      ERROR('two lists or two numbers expected',a,b);
>    fi;
>  end:
```

Test our procedure on the lists L1 and L2.

```
>  AND(L1,L2);
```

$$[1, 0, 1, 0, 1, 0, 0]$$

Loops and Truth Tables

One of the simplest uses of Maple is to test the validity of a particular proposition. For example, we might name a particular expression as

```
>  e1 := p or q;
```

$$e1 := p \text{ or } q$$

```
>  e2 := (not p) and (not q );
```

$$e2 := \text{not } (p \text{ or } q)$$

On input to Maple these simplify in such a way that it is obvious that not e1 and e2 will always have the same value no matter how p and q have been assigned truth values.

The implication "p implies q" is equivalent to "(not p) or q", and it is easy to write a Maple procedure to compute the latter.

```
>  implies := (p,q) -> (not p) or q;
```

$$implies := (p, q) \rightarrow \text{not } p \text{ or } q$$

To verify that this Maple definition of implies(p,q) is correct examine its value for all possible values of p and q.

```
>  implies(false,false), implies(false,true);
```

$$true, \, true$$

```
>  implies(true,false),  implies(true,true);
```
$$false, true$$

A systematic way of tabulating such truth values is to use the programming loop construct. Since much of what is computed inside a loop is hidden, we make use of the `print` statement to force selected information to be displayed. We can print out the value of p, q, and `implies(p,q)` in one statement as

```
>  print(p,q,implies(p,q));
```
$$p, q, \textbf{not } p \textbf{ or } q$$

To execute this print statement for every possible pair of values for `[p,q]` by placing one loop inside another.

```
>  for p in [false,true] do
>    for q in [false,true] do
>      print( p , q , implies(p,q) );
>    od:
>  od:
```
$$false, false, true$$
$$false, true, true$$
$$true, false, false$$
$$true, true, true$$

No matter how the `implies` truth values are computed, the "truth table" for the proposition "implies" must always have this structure.

This approach can be used to investigate many of the logical statements found in the supplementary exercises of this chapter. For example, the compound propositions such as found in Exercises 4 and 5 can be investigated as follows.

To verify that a proposition involving p and q is a tautology we need to verify that no matter what the truth value of p and the truth value of q, the proposition is always true. For example, To show that "((not q) and (p implies q)) implies (not q)" is a tautology we need to examine this proposition for all the possible truth value combinations of p and q. The proposition can be written as

```
>  p1 := implies( (not q) and implies(p,q) , not q );
```
$$p1 := true$$

For p **true**, and q **false**, the value of p1 is

```
>  subs( {p=true,q=false},p1);
```
$$true$$

The proposition p1 is completely described by its truth table.

```
> for p in [false,true] do
>   for q in [false,true] do
>     print(p,q,p1);
>   od;
> od;
```

$$false, false, true$$
$$false, true, true$$
$$true, false, true$$
$$true, true, true$$

When the variables p and q have been assigned values in the loop they retain that value until they are set to something else. Remember to remove such assignments by assigning p its own name as a value.

```
> p := 'p'; q := 'q';
```

$$p := p$$
$$q := q$$

We can generate a "truth table" for binary functions in exactly the same manner as we have for truth tables. Recall the definition of AND given in the previous section. A table of all possible values is given by:

```
> for i in [0,1] do
>   for j in [0,1] do
>     print(i,j,AND(i,j));
>   od:
> od:
```

$$0, 0, 0$$
$$0, 1, 0$$
$$1, 0, 0$$
$$1, 1, 1$$

We can even extend this definition of AND to one which handles pairs of numbers, or pairs of lists. The following procedure AND2 accomplishes this.

```
> AND2 := proc(a,b)
>   if not type([a,b],
>     {[numeric,numeric],[list(numeric),list(numeric)]})
>     then RETURN('AND2'(a,b));
>   fi;
>   AND(a,b);
> end:
```

Note that you can specify sets of types to type. As before, we have

```
> AND2(0,0); AND2([0,1],[0,0]);
```

$$0$$
$$[0, 0]$$

and when necessary, it can remain unevaluated as in

```
>  AND2(x,y);
```

$$AND2(x, y)$$

Comparing Two Propositions

Truth tables can be also be used to identify when two propositions are really equivalent.

A second proposition might be

```
>  p2 := p1 and q;
```

$$p2 := p1 \textbf{ and } q$$

To compare the truth tables for these two propositions (i.e. to test if they are equivalent) print out both values in a nested loop.

```
>  for p in [false,true] do
>    for q in [false,true] do
>      print(p,q,p1,p2);
>    od;
>  od;
```

$$false, false, p1, false$$
$$false, true, p1, p1$$
$$true, false, p1, false$$
$$true, true, p1, p1$$

How would you test if p2 was the same as p implies q?

Using Maple to Check Logical Arguments

This section show you how to use some of Maples logical operators to analyze "real life" logical arguments. We'll need to make use of some of the facilities in the logic package. The logic package is discussed in detail in Chapter 9. To load the logic package, we use the with command.

```
>  with(logic):
```

In particular, we shall require the bequal function, which tests for the logical equivalence of two logical (boolean) expressions. Procedures in the logic package operate upon boolean expressions composed with the *inert* boolean operators &and, &or, &not, and so on, in place of and, or, not. The inert operators are useful when you want to study the *form* of a boolean expression, rather than its value. Consult Chapter 9 for a more detailed discussion of these operators.

A common illogicism made in everyday life, particularly favored by politicians, is confusing the implication "a implies b" with the similar implication "not a implies not b". Maple has a special operator for representing the conditional operator '→'; it is &implies. Thus, we can see the following in Maple.

```
> bequal(a &implies b, &not a &or b);
```
$$true$$

Now, to see that '$a \to b$' and '$\bar{a} \to \bar{b}$' are *not* equivalent , and to further find particular values of a and b for which their putative equivalence fails, we can do the following.

```
> bequal(a &implies b, (&not a) &implies (&not b), 'assgn');
```
$$false$$

```
> assgn;
```
$$\{a = true, b = false\}$$

Another illogicism occurs when a conditional is confused with its converse. The **converse** of a conditional expression '$a \to b$' is the conditional expression '$b \to a$'. These are not logically equivalent.

```
> bequal(a &implies b, b &implies a, 'assgn');
```
$$false$$

```
> assgn;
```
$$\{a = true, b = false\}$$

However, a very useful logical principle is **contraposition**, which asserts that the implication '$a \to b$' is equivalent to the conditional '$\bar{b} \to \bar{a}$'. You can read this as: "a implies b" is equivalent to "not b implies not a", and you can prove it using Maple like this:

```
> bequal(a &implies b, &not b &implies &not a);
```
$$true$$

For more discussion of the logic package, and of the so-called "inert" operators, see Chapter 9.

1.2   Quantifiers and Propositions

Maple can be used to explore propositional functions and their quantification over a finite universe. To create a propositional function p such for which as "$p(x) = x > 0$" in Maple we enter

```
> p := (x) -> x > 0;
```
$$p := x \to 0 < x$$

The arrow notation `->` is really just an abbreviated notation for constructing the Maple procedure `proc(x) x>0 end`. Once defined, we can use p
to write propositions such as

> `p(x), p(3), p(-2) ;`

$$0 < x, 0 < 3, 0 < -2$$

To determine the truth value for specific values of x, we apply the `evalb`
procedure to the result produced by p. as in `evalb( p(3) )`.

We often wish to apply a function to every element of a list or a set. This
is accomplished in Maple by using the `map` command. The meaning of
the command `map(f,[1,2,3])` is best understood by trying it. To map
f onto the list `[1,2,3]`, use the command

> `map( f , [1,2,3] );`

$$[f(1), f(2), f(3)]$$

Each element of the list is treated, in turn, as an argument to f.

To compute the list of truth values for the list of propositions obtained
earlier, just use `map`.

> `map( evalb, [ p(x),p(3),p(-2)] );`

$$[-x < 0, true, false]$$

Note that the variable x has not yet been assigned a value, so the expression $p(x)$ does not yet simplify to a truth value.

Something similar can be done for multivariate propositional functions.

> `q := (x,y) -> x < y:`
> `evalb( q(3,0) );`

$$false$$

Maple can also be used to determine the truth value of quantified statements, provided that the universe of quantification is finite (or, at least,
can be finitely parameterized). In other words, Maple can be used to determine the truth value of such assertions as " for all x in S, $p(x)$", where
S is a *finite* set. For example, to test the truth value of the assertion: "For
each positive integer x less than or equal to 10, the inequality $100x > 2^x$
obtains." where set is

> `S := {seq(i, i = 1..10)}:`

first generate the set of propositions to be tested as

```
>  p := (x) -> 100*x > 2^x:
>  Sp := map( p , S );
```
$$Sp := \{2 < 100, 4 < 200, 8 < 300, 16 < 400, 32 < 500, 64 < 600,$$
$$128 < 700, 256 < 800, 512 < 900, 1024 < 1000\}$$

Next, compute the set of corresponding truth values.

```
>  Sb := map( evalb , Sp );
```
$$Sb := \{false, true\}$$

The quantified result is given by

```
>  if Sb = {true} then true else false fi;
```
$$false$$

A statement involving existential quantification, such as "there exists an x such that $p(x)$," is handled in much the same way, except that the resulting set of truth values have less stringent conditions to satisfy. For example, to test the truth value of the assertion: "There is a positive integer x not exceeding 10 for which $x^2 - 5$ is divisible by 11". over the same universe of discourse as before (the set S of positive integers less than or equal to 10) construct the set of propositions and their truth values as before

```
>  q := (x) -> (irem(x^2 - 5, 11) = 0):
>  Sp := map(q,S);
```
$$Sp := \{-4 = 0, -1 = 0, 4 = 0, 0 = 0, 9 = 0, 10 = 0, 7 = 0\}$$

```
>  Sb := map(evalb,Sp);
```
$$Sb := \{false, true\}$$

The `irem` procedure returns the integral remainder upon division of its first argument by its second. The existential test is just

```
>  if has( Sb , true ) then true else false fi;
```
$$true$$

To test different propositions, all you need do is change the universe S and the propositional function p.

If the universe of discourse is a set of ordered pairs, we can define the propositional function in terms of a list. For example, the function

```
>  q := (vals::list) -> vals[1] < vals[2]:
```

evaluates as

```
>  q( [1,30] );
```
$$1 < 30$$

A set of ordered pairs can be constructed using nested loops. To create the set of all ordered pairs (a, b) from the two sets, A and B use nested loops as in

```
> A := {1,2,3}: B := {30,60}: S := NULL:
> for a in A do
>   for b in B do
>     S := S , [a,b];
>   od:
> od:
```

The desired set is

```
> {S};
```

$$\{[1, 30], [1, 60], [2, 30], [2, 60], [3, 30], [3, 60]\}$$

1.3   Sets

As we have seen in the earlier sections, sets are fundamental to the description of almost all of the discrete objects that we study in this course. They are also fundamental to Maple. As such, Maple provides extensive support for both their representation and manipulation.

Maple uses curly braces ({ , }) to represent sets. The empty set is just

```
> {};
```

$$\{\}$$

A Maple set may contain any of the objects known to Maple. Typical examples are shown here.

```
> {1,2,3};
```

$$\{1, 2, 3\}$$

```
> {a,b,c};
```

$$\{a, b, c\}$$

One of the most useful commands for constructing sets or lists is the **seq** command. For example, to construct a set of squares modulo 57, you can first generate a sequence of the squares as in

```
> s1 := seq( i^2 mod 30,i=1..60);
```

$$s1 := 1, 4, 9, 16, 25, 6, 19, 4, 21, 10, 1, 24, 19, 16, 15, 16, 19, 24, 1, 10,$$
$$21, 4, 19, 6, 25, 16, 9, 4, 1, 0, 1, 4, 9, 16, 25, 6, 19, 4, 21, 10, 1, 24, 19,$$
$$16, 15, 16, 19, 24, 1, 10, 21, 4, 19, 6, 25, 16, 9, 4, 1, 0$$

This can be turned into a set by typing

```
>  s2 := {s1};
```

$$s2 := \{0, 1, 4, 6, 9, 10, 15, 16, 19, 21, 24, 25\}$$

Note that there are no repeated elements in the set s2. An interesting example is:

```
>  {seq(randpoly(x,degree=2),i=1..5)};
```

$$\{45\,x^2 - 8\,x - 93, -85\,x^2 - 55\,x - 37, -35\,x^2 + 97\,x + 50,$$
$$79\,x^2 + 56\,x + 49, 63\,x^2 + 57\,x - 59\}$$

The randpoly procedure creates a random polynomial of degree equal to that specified with the degree option (here 2). Thus, the last example above has generated a set consisting of 5 random quadratic polynomials in the indeterminate x.

The ordering of the elements is not always the same as the order you used when you defined the set. This is because Maple displays members of a set in the order that they are stored in memory (which is not predictable). By definition, the elements of a set do not appear in any particular order, and Maple takes full advantage of this to organize its storage of the sets and their elements in such a way that comparisons are easy for Maple. This can have some surprising consequences. In particular, you cannot sort a set. Use lists instead. If order is important, or if repeated elements are involved, use lists. Simple examples of the use of lists include

```
>  r := rand(100): # random no. < 100
>  L := [seq(r(), i=1..20)]; # list of 20 random nos. < 100
```

$$L :=$$
$$[85, 16, 61, 7, 49, 86, 98, 66, 9, 73, 81, 74, 66, 73, 42, 91, 93, 0, 11, 38]$$

```
>  N := [1,1,1,2,2,2,3,3,3];
```

$$N := [1, 1, 1, 2, 2, 2, 3, 3, 3]$$

Such a lists can be sorted using the sort command.

```
>  M := sort(L);
```

$$M :=$$
$$[0, 7, 9, 11, 16, 38, 42, 49, 61, 66, 66, 73, 73, 74, 81, 85, 86, 91, 93, 98]$$

The sort procedure sorts the list L producing the list M in increasing numerical order.

The number of elements in N is:

```
>  nops(N);
```

To find out how many distinct elements there are in a list simply convert it to a set, and compare the size of the set to the size of the original list by using the command `nops`.

```
>  NS := convert(N,set);
```
$$NS := \{1, 2, 3\}$$

```
>  nops(NS);
```
$$3$$

Maple always simplifies sets by removing repeated elements and reordering the elements to match its internal order. This is done to make it easier for Maple to compute comparisons.

To *test* for equality of two sets write the set equation $A = B$ and can force a comparison using the `evalb` command.

```
>  A = B;
```
$$A = B$$

```
>  evalb( A = B );
```
$$false$$

Set Operations

Given the two sets

```
>  A := {1,2,3,4}; B := {2,1,3,2,2,5};
```
$$A := \{1, 2, 3, 4\}$$
$$B := \{1, 2, 3, 5\}$$

We can compute the relative difference of two sets using `minus`, as in

```
>  A minus B;
```
$$\{4\}$$

```
>  B minus A;
```
$$\{5\}$$

We can also construct their union

```
>  C := A union B;
```
$$C := \{1, 2, 3, 4, 5\}$$

Several other set operations are supported in Maple. For instance, you can determine the power set of a given finite set using the `powerset` command from the `combinat` package. To avoid having to use the `with` command to load the entire `combinat` package, you can use its "full name" as follows.

```
>  S := {1,2,3}:
>  pow_set_S := combinat[powerset](S);
```
$$pow\_set\_S := \{\{\}, \{1, 2, 3\}, \{2, 3\}, \{3\}, \{1, 3\}, \{1\}, \{2\}, \{1, 2\}\}$$

Try this with some larger sets.

The symmetric difference operator `symmdiff` is used to compute the symmetric difference of two or more sets. You will need to issue the command

```
>  readlib(symmdiff):
```

before you can use `symmdiff`. Then, the symmetric difference of A and B is

```
>  symmdiff(A, B);
```
$$\{4, 5\}$$

Recall that the symmetric difference of two sets A and B is defined to be the set $(A \cup B) - (A \cap B)$ of objects that belong to exactly one of A and B.

```
>  symmdiff(A, B);
```
$$\{4, 5\}$$

```
>  (A union B) minus (A intersect B);
```
$$\{4, 5\}$$

To construct the Cartesian product of two sets, we write a little procedure in Maple as follows. This procedure will construct the Cartesian product $A \times B$ of the two sets A and B given to it as arguments.

```
>  CartesianProduct := proc(A::set, B::set)
>     local prod, # the Cartesian product; returned
>           a,b; # loop variables
>
>     prod := NULL; # initialize to a NULL sequence
>     # loop like crazy
>     for a in A do
>       for b in B do
>         # add the ordered pair [a,b] to the end
>         prod := prod, [a,b];
>       od;
>     od;
>     RETURN({prod}); # return a set
>  end:
```

The procedure is called by providing it with two sets as arguments.

```
>  S := {1,2,3,4};
```
$$S := \{1, 2, 3, 4\}$$

```
>  T := {'Bill', 'Hillary', 'Chelsea', 'Socks'};
```
$$T := \{\mathit{Bill}, \mathit{Hillary}, \mathit{Chelsea}, \mathit{Socks}\}$$

```
>  P := CartesianProduct(S, T);
```
$$P := \{[1, \mathit{Bill}], [1, \mathit{Hillary}], [1, \mathit{Chelsea}], [1, \mathit{Socks}], [2, \mathit{Bill}], [2, \mathit{Hillary}],$$
$$[2, \mathit{Chelsea}], [2, \mathit{Socks}], [3, \mathit{Bill}], [3, \mathit{Hillary}], [3, \mathit{Chelsea}], [3, \mathit{Socks}],$$
$$[4, \mathit{Bill}], [4, \mathit{Hillary}], [4, \mathit{Chelsea}], [4, \mathit{Socks}]\}$$

Note that the order in which the arguments appear is relevant.

```
>  Q := CartesianProduct(T, S);
```
$$Q := \{[\mathit{Bill}, 1], [\mathit{Bill}, 2], [\mathit{Bill}, 3], [\mathit{Bill}, 4], [\mathit{Hillary}, 1], [\mathit{Hillary}, 2],$$
$$[\mathit{Hillary}, 3], [\mathit{Hillary}, 4], [\mathit{Chelsea}, 1], [\mathit{Chelsea}, 2], [\mathit{Chelsea}, 3],$$
$$[\mathit{Chelsea}, 4], [\mathit{Socks}, 1], [\mathit{Socks}, 2], [\mathit{Socks}, 3], [\mathit{Socks}, 4]\}$$

The representation and manipulation of infinite sets is somewhat more complicated. Discussion of this topic will occur in Chapter 10.

New sets and lists can also be created by mapping functions onto them. For example, you can map an unknown function onto a set, as in

```
>  s3 := map(f,s2);
```
$$s3 :=$$
$$\{f(10), f(15), f(16), f(19), f(21), f(24), f(25), f(0), f(1), f(6), f(9), f(4)\}$$

Note that the ordering of the elements in s3 need not have any relationship with the ordering of the elements in s2. Both are sets and order is irrelevant.

It may happen that f requires a second argument. If so, map can still be used as:

```
>  map(f, s2, y);
```
$$\{f(0, y), f(4, y), f(6, y), f(9, y), f(10, y), f(1, y), f(15, y), f(16, y), f(19, y),$$
$$f(21, y), f(24, y), f(25, y)\}$$

Again, the ordering is irrelevant, and in this case, because f is undefined, the result shows you explicitly what map has done.

You can also map onto lists. For example, given the list

```
>  l2 := convert(s2,list);
```
$$l2 := [0, 1, 4, 6, 9, 10, 15, 16, 19, 21, 24, 25]$$

the list (in their correct order) of remainders of these numbers on division
by 6 is just

```
>  map( modp , 12 , 6 );
```
$$[0, 1, 4, 0, 3, 4, 3, 4, 1, 3, 0, 1]$$

where modp is a two argument procedure used to calculate remainder on
division, as in

```
>  modp(23,6);
```

5

1.4   Functions and Maple

For a discussion of the concept of mathematical functions see section 1.6
of the main text book. Functions are supported by Maple in a variety of
ways. The two most direct constructs are tables and procedures.

Tables

Tables can be used to define functions when the domain is finite and
relatively small. To define a function using a table we must associate to
each element of the domain an element of the codomain of this function.

A table t defining such a relationship can be defined by the command

```
>  t := table([a=a1,b=b1,c=c1]);
```
$$t := \text{table}([$$
$$b = b1$$
$$a = a1$$
$$c = c1$$
$$])$$

Once the table t is defined in this manner, the values of the expressions
t[a], t[b] and t[c] are

```
>  t[a];
```
$$a1$$

```
>  t[b];
```
$$b1$$

```
>  t[c];
```
$$c1$$

The set of entries $\{a, b, c\}$ occurring inside the square brackets form the domain of this discrete function. They are called *indices* in Maple. They can be found by using the `indices` command. For example, the *set* of indices of t is

```
>  idx := { indices(t) };
```
$$idx := \{[b], [a], [c]\}$$

Each *index* is presented as a list. This is to allow for very complicated indices, perhaps involving pairs or triples such as t[x,y,z].

In cases such as the above where the indices are simply names, the set of names can be recovered by applying a Maple procedure to every element of the set. Since

```
>  op( [a] );
```
$$a$$

evaluates to the single element contained inside this single element list [a], we can recover the set of names by using the `map` command to apply the `op` command to every element of the set idx. This required command is:

```
>  map( op , idx );
```
$$\{c, b, a\}$$

The set $\{t[a], t[b], t[c]\}$ constitutes the *range* of the discrete function and can be recovered by the command `entries`, which returns the sequence of entries from a table, each represented as a list. To compute the range of a function represented by the table t, you can use `map` and `op` as before.

```
>  rng := map(op, {entries(t)});
```
$$rng := \{c1, b1, a1\}$$

The number of elements in the domain is just

```
>  nops(idx);
```
$$3$$

The number of elements in the range is just

```
>  nops(rng);
```
$$3$$

Adding New Elements

To add a new element to a table, simply use the assignment operator.

```
>  t[d]  := d1;
```

$$t_d := d1$$

Use the commands `indices` and `entries` to verify that this has extended the definition of t.

```
>  indices(t);
```

$$[d], [b], [a], [c]$$

```
>  entries(t);
```

$$[d1], [b1], [a1], [c1]$$

Tables versus Table Elements

You can refer to a table by either its name as in

```
>  t;
```

$$t$$

or its value as in

```
>  eval(t);
```

$$\begin{aligned} \text{table}([\\ d = d1 \\ b = b1 \\ a = a1 \\ c = c1 \\ ]) \end{aligned}$$

This crucial distinction is made because tables can have thousands of elements. It allows you to focus on the table as a single entity (represented by a name), or looking at all the detail through the elements themselves.

Defining Functions via Rules

Not all relations or functions are defined over finite sets. Often, in non-finite cases, the function is defined by a rule associating elements of the domain with elements of the range.

Maple is well suited for defining functions via rules. Simple rules (such as $x \longrightarrow x^2 + 3$) can be specified using Maple's "$\longrightarrow$" operator as in

```
>  (x) -> x^2 + 3;
```

$$x \rightarrow x^2 + 3$$

Such a rules are very much like other Maple objects. They can be named, or re-used to form other expressions. For example, we name the above rule as f by the assignment statement

```
>  f := (x) -> x^2 + 3;
```
$$f := x \rightarrow x^2 + 3$$

To use such a rule, we "apply" it to an element of the domain. The result is an element of the range. Examples of function application are:

```
>  f(3);
```
$$12$$

```
>  f(1101101);
```
$$1212423412204$$

We can even apply functions to indeterminates such as t as in

```
>  f(t);
```
$$t^2 + 3$$

The result is a formula which is dependent on t.

```
>  f(t);
```
$$t^2 + 3$$

You can even use an undefined symbol g as if it were a rule. Because the rule is not specified, the result returns unevaluated and in a form which can evaluate at some later time after you have defined a suitable rule. Examples of this include:

```
>  g(3);
```
$$g(3)$$

```
>  g(t);
```
$$g(t)$$

The ordered pair describing the effect of a function g on the domain element t is just:

```
>  [t,g(t)];
```
$$[t, g(t)]$$

An algebra of functions

Just as for tables, functions can be manipulated by name or value. To see the current definition of a function use the `eval()` command, as in

```
>  eval(f);
```

$$x \to x^2 + 3$$

If there is no rule associated with the name then the result will be a name.

```
>  eval(g);
```

$$g$$

Depending on how your Maple session is configured, you may need to issue the command

```
>  interface(verboseproc=2);
```

and then re-evaluate `eval(f)` before seeing the details of the function definition. The `verboseproc` parameter controls how much information is displayed when an expression is evaluated. It is primarily used to view the source code for Maple library procedures, as shown above, but can be used to view the code for user functions, as well.

One advantage of being able to refer to functions by name only is that you can create new functions from old ones simply by manipulating them algebraically. For example the algebraic expressions

```
>  f + g;
```

$$f + g$$

```
>  g^2;
```

$$g^2$$

and

```
>  h := f^2;
```

$$h := f^2$$

each represent functions. To discover the rule corresponding to each of these new function definitions, simply apply them to an indeterminate. For these examples, we obtain

```
>  (f + g)(t);
```

$$t^2 + 3 + g(t)$$

```
>  (g^2)(t);
```

$$g(t)^2$$

and

```
>  h(t);
```

$$(t^2 + 3)^2$$

Notice that in each case presented here, **g** is undefined so that **g(t)** is the algebraic expression that represents the result of applying the function **g** to the indeterminate **t**.

Even numerical quantities can represent functions. The rule

```
>  one := (x) -> 1;
```

$$one := 1$$

simplifies to 1 and always evaluates to 1 when applied as a function.

```
>  one(t);
```

$$1$$

The result is that

```
>  (g +1)(t);
```

$$g(t) + 1$$

behaves exactly as if **1** were a function name for the function $(x) \longrightarrow 1$. This generalizes to all numeric quantities. In particular **(3)*(x)** and **(3)(x)** behave very differently as the first one is multiplication and the second one is an application of the constant function $(x) \longrightarrow 3$

```
>  (3)*x, (3)(x);
```

$$3\,x, 3$$

In both cases the parenthesis can be left off of the **3** with no change in the outcome.

Functional Composition

Maple uses the **@** operator to denote functional composition. The composition $f \cdot g$ is entered in Maple as **f@g**. In a new Maple session the outcome of applying the function **h = f@g** to **t** is

```
>  restart;
>  h := f@g;
```

$$h := f@g$$

```
>  h(t);
```

$$f(g(t))$$

Functions may be composed with themselves as in

```
>   g := f@@3;
```

$$g := f^{(3)}$$

The parenthesis around the exponent are important. They indicate that composition rather than multiplication is taking place. Again, this meaning becomes clear if you apply the function **g** to an unknown **t**, as in

```
>   g(t);
```

$$f^{(3)}(t)$$

Constructing Functional Inverses

The identity function **Id** is the function

```
>   Id := (x) -> x;
```

$$Id := x \to x$$

A functional inverse of the function f is a function g that when composed with f results in the identity function $(x) \longrightarrow x$. For example, given

```
>   f := (x) -> 5*x^3 + 3;
```

$$f := x \to 5\,x^3 + 3$$

the inverse of **f** is a function **g** such that

```
>   (f@g)(t) = t;
```

$$5\,g(t)^3 + 3 = t$$

Use this equation to deduce that **g(t)** should be

```
>   isolate(",g(t));
```

$$g(t) = \text{RootOf}(5\,\_Z^3 - t + 3)$$

The right hand side of this equation can be used to actually define the function g. The Maple **unapply()** command can be used to turn the expression into a function. The righthand side is:

```
>   rhs(");
```

$$\text{RootOf}(5\,\_Z^3 - t + 3)$$

To turn this expression into a rule, use **unapply()**, specifying the name used in the general rule as an extra argument.

```
>   g := unapply(",t);
```

$$g := t \to \text{RootOf}(5\,\_Z^3 - t + 3)$$

In this example, the resulting function is named **g**.

1.5   Growth of Functions

The primary tool used to study growth will be plotting. This is handled in Maple by means of is shown here.

```
>  plot( {ln(x),n, n*ln(n) } , n=2..6 );
```

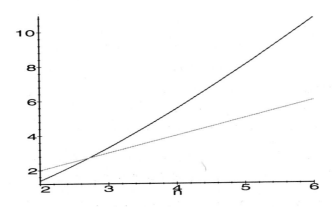

For a single curve, omit the set braces, as in

```
>  plot( x^2 + 3 , x = 0..4 , y = 0..10);
```

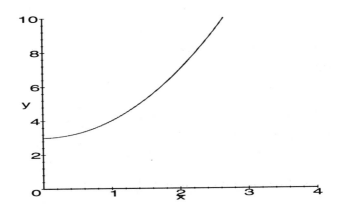

The first argument to the plot command specifies the function or curve, or a set of curves. The second argument specifies a domain, while the optional third argument $y = a..b$ specifies a range. See the plots package for a wide variety of additional commands. Also, see the help page for "plot,options".

It is possible to plot functions that are not continuous, provided that they are at least piecewise continuous. Two good examples relevant to discrete mathematics are the `floor` and `ceil` ("ceiling") functions. Here, we plot both on the same set of axes.

```
>  plot({floor(x), ceil(x)}, x = -10..10);
```

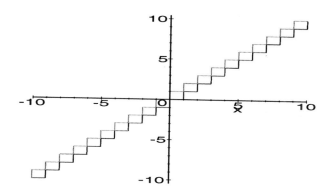

1.6   Computations and Explorations

1. What is the largest value of n for which $n!$ has fewer than 100 decimal digits and fewer than 1000 decimal digits?

   Solution: The number of digits in a decimal integer can be determined in Maple by using the `length` function.

   ```
   >  length(365);
   ```

   $$3$$

   To answer this question we can use the `length` function to construct a test for a `while` loop.

   ```
   >  n := 1;
   ```

   $$n := 1$$

   ```
   >  while length(n!) < 10 do
   >    n := n + 1;
   >  od:
   >  n - 1;
   ```

   $$12$$

   ```
   >  length((n - 1)!);
   ```

   $$9$$

The same technique will allow you to find the largest n for which $n!$ has fewer than 100 or fewer than 1000 digits.

3. Calculate the number of one to one functions from a set S to a set T, where S and T are finite sets of various sizes. Can you determine a formula for the number of such functions? (We will find such a formula in Chapter 4.)

Solution: We'll show here how to count the number of one to one functions from one finite set to another and leave the conjecturing to the reader. Since the number of one to one functions from one set to another depends only upon the sizes of the two sets, we may as well use sets of integers. A one to one function from a set S to a set T amounts to a choice of $|S|$ members of T and a permutation of those elements. Suppose that S has 3 members, we can view a one to one function from S to T as a labeling of 3 members of T with the members of S. That is we wish to choose 3 members of T and then permute them in all possible ways. We can compute these permutations with the function `permute` in the `combinat` package. If we assume that T has 5 members, then we enter the command

```
>  combinat[permute](5, 3);
```
$$[[1,2,3],[1,2,4],[1,2,5],[1,3,2],[1,3,4],[1,3,5],[1,4,2],[1,4,3],$$
$$[1,4,5],[1,5,2],[1,5,3],[1,5,4],[2,1,3],[2,1,4],[2,1,5],[2,3,1],$$
$$[2,3,4],[2,3,5],[2,4,1],[2,4,3],[2,4,5],[2,5,1],[2,5,3],[2,5,4],$$
$$[3,1,2],[3,1,4],[3,1,5],[3,2,1],[3,2,4],[3,2,5],[3,4,1],[3,4,2],$$
$$[3,4,5],[3,5,1],[3,5,2],[3,5,4],[4,1,2],[4,1,3],[4,1,5],[4,2,1],$$
$$[4,2,3],[4,2,5],[4,3,1],[4,3,2],[4,3,5],[4,5,1],[4,5,2],[4,5,3],$$
$$[5,1,2],[5,1,3],[5,1,4],[5,2,1],[5,2,3],[5,2,4],[5,3,1],[5,3,2],$$
$$[5,3,4],[5,4,1],[5,4,2],[5,4,3]]$$

The first arguments (here 5, the size of T) is the number of objects that we want to permute, and the (optional) second argument is the number of elements to permute. To count them, we use the `nops` routine.

```
>  nops(");
```
$$60$$

If, instead, the set S had, say, 4 members, then we would compute:

```
>  nops(combinat[permute](5, 4));
```
$$120$$

5. We know that n^b is $O(d^n)$ when b and d are positive numbers with $d \geq 2$. Give values of the constants C and k such that $n^b \leq Cd^n$ whenever $n > k$ for each of the following sets of values: $b = 10$, $d = 2$; $b = 20$, $d = 3$; $b = 1000$, $d = 7$.

Solution: Here we solve only the last of these, leaving the rest for the reader. We are seeking values of the constants C and k such that $n^{1000} \leq C \cdot 7^n$ whenever $n > k$. We'll substitute a test value for C, and then use a loop to test for which values of n the inequality is satisfied.

```
>  n^1000 < C * 7^n;
```
$$n^{1000} < C\, 7^n$$

```
>  left := lhs(");
```
$$left := n^{1000}$$

```
>  right := rhs("");
```
$$right := C\, 7^n$$

```
>  right_sub := subs(C = 2, right);
```
$$right\_sub := 2\, 7^n$$

```
>  k := 1;
```
$$k := 1$$

```
>  while evalb(subs(n = k,left) >= subs(n = k, right_sub)) do
>     k := k +1 ;
>  od;
```

You should also try this for other test values of C.

You can also try to solve the equation $n^{1000} = C * 7^n$ for n, using Maples `solve` routine. For this particular example, you will need to use the help facility to learn more about the W function, which satisfies the equation $W(x)e^{W(x)} = x$. (It is a complex valued function, but is real valued for $x > 1/e$.)

1.7   Exercises/Projects

Exercise 1. Use computation to discover what the largest value of n is for which $n!$ has fewer than 1000 digits.

Exercise 2. Compare the rate of growth of factorials and the function f defined by $(x) \to e^{2x}$.

Exercise 3. We saw that a list T with four elements had $24 = 4!$ permutations. Does this relationship hold true for smaller sets T? Go back and change the list T and re-compute the subsequent values. Does this relationship hold true for larger lists (say of size 5 or 6)? (Be careful as the number $n!$ grows very rapidly!)

Exercise 4. Can you conjecture what the answer would be for larger n? Can you prove your conjecture? We will construct such a formula in Chapter 4.

Exercise 5. Develop Maple procedures for working with fuzzy sets, including procedures for finding the complement of a fuzzy set, the union of fuzzy sets, and the intersection of fuzzy sets. (See Page 58 of the text.)

Exercise 6. Develop Maple procedures for finding the truth value of expressions in fuzzy logic. (See Page 13 of the text.)

Exercise 7. Develop maple procedures for working with multisets. In particular, develop procedures for finding the union, intersection, difference, and sum of two multisets. (See Page 57 of the text.)

2 The Fundamentals: Algorithms, Integers and Matrices

This chapter covers material related to algorithms, integers and matrices. An *algorithm* is a definite procedure that solves a problem in a finite number number of steps. In Maple, we implement algorithms using *procedures* that take input, process this input and output desired information. Maple offers a wide range of constructs for looping, condition testing, input and output that allows almost any possible algorithm to be implemented.

We shall see how to use Maple to study the complexity of algorithms. In particular, we shall describe several ways to study the time required to perform computations using Maple.

The study of integers, or number theory, can be pursued using Maple's `numtheory` package. This package contains a wide range of functions that do computations in number theory, including functions for factoring, primality testing, modular arithmetic, solving congruences, and so on. We will study these and other aspects of number theory in this chapter.

Maple offers a complete range of operations to manipulate and operate on matrices, including all the capabilities discussed in the chapter of the text. In this chapter we will only touch on Maple's capabilities for matrix computations. In particular we will examine matrix addition, multiplication, transposition and symmetry, as well as the meet, join and product operations for Boolean matrices.

2.1 Implementing Algorithms in Maple

When creating algorithms, our goal is to determine a finite sequence of actions that will perform a specific action or event. We have already seen numerous examples of procedures or algorithms written in
Maple. This chapter will provide further examples and touch on some of the built in procedures and functions that Maple provides that can make your task easier.

The following simple example serves to illustrate the general syntax of a procedure in
Maple.

```
>   Alg1 := proc(x::numeric)
>     global a; local b;
>     a := a + 1;
>     if x < 1 then b := 1;
>     else b := 2; fi;
>     a := a + b + 10;
>     RETURN(a);
>   end:
```

A procedure definition begins with the key word **proc** and ends with the key word **end**. The bracketed expression (x::numeric) immediately following **proc** indicates that one argument is expected whose value is to be used in place of the name x in any computations which take place during the execution of the procedure. The type **numeric** indicates that value that is provided during execution must be of type **numeric**. Type checking is optional.

The statement **global a;** indicates that the variable named a is to be borrowed from the main session in which the procedure is used. Any changes that are made to its value will remain in effect after the procedure has finished executing. The statement **local b;** indicates that one variable named b is to be *local* to the procedure. Any values that the variable b takes on during the execution of the procedure disappear after the procedure finishes. The rest of the procedure through to the **end** (the procedure body), is a sequence of statements or instructions that are to be carried out during the execution of the procedure. Just like any other object in Maple, a procedure can be assigned to a name and a colon can be used to suppress the display of the output of the assignment statement.

Procedure Execution

Prior to using a procedure, you may have assigned values to the variables a and b, as in

```
>   a := x^2 ; b := 10;
```

$$a := x^2$$
$$b := 10$$

The procedure is invoked (executed) by following its name with a parenthasized list of arguments. The argument cannot be anything other than a number. Otherwise an error will occur as in

```
>   Alg1(z);
```

```
Error, Alg1 expects its 1st argument, x,
to be of type numeric, but received z
```

For a numeric argument, the algorithm proceeds to execute the body of the procedure, as in.

> `Alg1(1.3);`

$$x^2 + 13$$

Execution proceeds (essentially) as if you had replaced every occurrence of x in the body of the procedure by (in this case) 1.3 and then then executed the body statements one at a time, in order. Local variables have no value initially, while global variables have the value they had before starting execution unless they get assigned a new value during execution.

Execution finishes when you get to the last statement, or when you encounter a RETURN statement, which ever occurs first. The value returned by the procedure is either the last computed value or the value indicated in the RETURN statement. You may save the returned value by assigning it to a name, as in

> `result := Alg1(1,3);`

$$result := x^2 + 26$$

or you may use it in further computations, as in

> `Alg1(1,3) + 3;`

$$x^2 + 42$$

It may happen that the procedure does not return any value. In fact, this can be done deliberately by executing the special statement RETURN(NULL). This would be done if, for example, the procedure existed only to print a message.

Local and Global Variables

What happens to the values of a and b?

Since a was global, any changes to its value that took place while the procedure was executing remain in effect. To see this, observe that the value of a has increased by 1.

> `a;`

$$x^2 + 39$$

The variable b was declared local to the procedure body, so even though its value changed during execution, those changes have no effect on the value of the global variable b. To see this, observe that the global value of b remains unchanged at

```
>  b;
```

$$10$$

Local variables exist to assist with the actions that take place during execution of the procedure. They have no meaning outside of that computation.

During the execution of the body of the procedure, assignments are made, and the sequence of actions is decided by the use of two main control structures, loops and conditional branches.

A loop is a giant single statement which may have other statements inside it. Loops come in many forms. Two of the most common forms appear in the following sample procedures for printing numbers up to a given integer value.

The `for` loop typically appears as in

```
>  MyForLoop := proc(x::integer)
>     local i;
>     for i from 1 to x do
>        print(i);
>     od;
>  end:
```

The result of executing this procedure is

```
>  MyForLoop(3);
```

$$1$$
$$2$$
$$3$$

The `while` loop typically appears as in

```
>  MyWhileLoop := proc(x::integer)
>     local j;
>     j := 1;
>     while j < x do
>        print(j);
>        j := j + 1;
>     od;
>  end:
```

The result of executing this procedure is

```
>  MyWhileLoop(3);
```

$$1$$
$$2$$

3

In both cases, the statements that are repeated are those found between the two key words do and od.

Conditional statements (based on the if statement) also come in several forms, the most common of which is illustrated in the procedure definition below. They form a single giant statement, each part of which may have its own sequence of statements. This functionality allows procedures to make decisions based on true or false conditions.

```
>   DecisionAlg := proc(y::integer)
>     if (y< 5) then
>       print('The input is less than 5');
>     elif (y = 5) then
>       print('The number is equal to 5');
>     else
>       print('The number is larger than 5');
>   fi;
>   end:
```

The outcome differs depending on what the argument value is at the time the procedure is invoked.

```
>   DecisionAlg(10);
```
$$\textit{The number is larger than 5}$$

```
>   DecisionAlg(5);
```
$$\textit{The number is equal to 5}$$

```
>   DecisionAlg(-1001);
```
$$\textit{The input is less than 5}$$

This basic conditional statement forms the basis for decision making or *branching* in Maple procedures.

We now will combine aspects of all three of these programming tools to build a procedure directly from the problem statement stage, through the *pseudocode*, to the final Maple code.

Consider the following problem: *Given an array of integers, find and output the maximum and minimum elements.* So, an algorithm for solving this problem will take the form of inputing an array of integers, processing the array in some manner to extract the maximum and minimum elements, and then outputting these two values. Now, let's take what we have just outlined and make it more rigorous. That is, we wish to form *pseudocode*, which is not written in any specific computer language but allows easy translation into (almost) any computer language. In this case, we shall go point by point over the steps of our algorithm, called

MaxAndMin. Again, the algorithm steps are constructed in a logical manner as follows:

1. The array is given as input. We call this input t.

2. We set the largest element and smallest element equal to the first element of the array.

3. We loop through the entire array, element by element. We call the current position cur_pos.

4. If the current element at cur_pos in the array is larger than our current maximum, we replace our current maximum with this new maximum.

5. If the element at cur_pos in the array is smaller than our current minimum, we replace our current minimum with this new minimum.

6. Once we reach the end of the array, we have compared all possible elements, so we output our current minimum and maximum values. They must be the largest and smallest elements for the entire array, since we have scanned the entire array.

Now, we convert each line of our pseudocode into Maple syntax. The reader should notice that each line of pseudocode translates almost directly into Maple syntax, with the keywords of the pseudocode line being the keywords of the Maple line of code.

To use the array functionality of Maple, we need to first load the linalg package. This loading outputs two warnings, which indicate that the two previous defined Maple functions for norm and trace have been overwritten with new definitions. Since these two functions will not be used in the following example, we can ignore the warnings and proceed with the procedure implementation.

```
>   with(linalg):
>   MaxAndMin := proc(t::array)
>     local cur_max, cur_min, cur_pos;
>     cur_max := t[1];
>     cur_min := t[1];
>     for cur_pos from 1 to vectdim(t) do
>       if t[cur_pos] > cur_max then
>         cur_max := t[cur_pos]
>       fi;
>       if t[cur_pos] < cur_min then
>         cur_min := t[cur_pos]
>       fi;
```

```
>    od;
>    RETURN([cur_min, cur_max]);
>  end:
```

We show the output of this procedure on two arrays of integers.

```
>  t := array(1..6, [1, 2, 45, 3, 2,10]);
```
$$t := [1, 2, 45, 3, 2, 10]$$

```
>  r := array(1..5, [5, 4, 9, 10, 16]);
```
$$r := [5, 4, 9, 10, 16]$$

```
>  MaxAndMin(t);
```
$$[1, 45]$$

```
>  MaxAndMin(r);
```
$$[4, 16]$$

This example shows that the steps from pseudocode to Maple code are straightforward and relatively simple. However, keep in mind, many of these types of operations are already available as part of the Maple library. For example, the maximum of an array could be computed as in

```
>  tlist := convert( eval(t), list ):
>  max( op(tlist) );
```
$$45$$

2.2 Measuring the Time Complexity of Algorithms in Maple

We are interested not only in the accuracy and correctness of the algorithms that we write, but also in their speed, or efficiency. Often, we are able to choose from among several algorithms that correctly solve a given problem. However, some algorithms for solving a problem may be more efficient than others. To choose an algorithm wisely requires that we analyze the efficiency of the various choices before us. This must be done in two ways: first, a mathematical analysis of the algorithm must be carried out, to determine its average and worst case running time; second, a practical implementation of the algorithm must be written, and tests made to confirm the theory. Maple cannot do the mathematical analysis for you, but it does provide several facilities for measuring the performance of your code. We shall discuss these facilities in this section.

First, note that Maple offers a way to measure the specific CPU (Central Processing Unit) time that a function used to compute a result. This is illustrated in the following example.

```
>  st := time():
>  MaxAndMin(r): MaxAndMin(t): MaxAndMin(t):
>  time()-st;
```
$$0$$

The `time` procedure reports the total number of seconds that have elapsed during the current Maple session. Here, we record the start time in the variable `st`, run the procedures that we wish to time, and then compute the time difference by calculating `time() - st`. This gives the time required to run the commands in seconds.

To illustrate this `time` function further, we will write a new `ManyFunctions` procedure that carries out some computations, but does not print any output. The reason is that our test case for output would normally output approximately 2 pages of digits, and this is not of interest to us here.

```
>  ManyFunctions := proc(x)
>     local a,b,c,d,e;
>     a := x;
>     b := x^2;
>     c := x^3;
>     d := x!;
>     e := x^x;
>  end:
>  st := time():
>  ManyFunctions(1000):
>  time() - st;
```
$$.067$$

This standard technique for timing computations will be used occasionally throughout the remainder of the book.

Also, Maple allows use to keep track of any additions, multiplications and functions that we may wish to use, by way of the `cost` function. The following example illustrates its usage.

```
>  readlib(cost):
>  cost(a^4 + b + c + (d!)^4 + e^e);
```
$$5 \ additions + 8 \ multiplications + 3 \ functions$$

We use the `readlib` command to load the definition of the `cost` function into the current Maple session. This is necessary for some of Maple's library routines, but not for many. The help page for a particular function should tell you whether or not you need to load the definition for that function with a call to `readlib`.

So, the `cost` and `time` functions help measure the complexity a given procedure. Specifically, we can analyze the entire running time for a procedure by using the `time` command and we can analyze a specific line of code by using the `cost` command to examine the computation costs in terms of multiplications, additions and function calls required to execute that specific line of Maple code.

We will now use these functions to compare two algorithms that compute the value of a polynomial at a specific point. We would like to determine which algorithm is faster for different inputs to provide some guidance as to which is more practical. To begin this analysis, we construct procedures that implement the two algorithms, which are outlined in pseudocode on Page 110 of the textbook.

```
> Polynomial := proc(c::float, coeff::list)
>   local power, i,y;
>   power := 1;
>   y := coeff[1];
>   for i from 2 to nops(coeff) do
>     power := power*c;
>     y := y + coeff[i] * power;
>   od;
>   RETURN(y);
> end:
> Horner := proc(c::float, coeff::list)
>   local power, i,y;
>   y := coeff[nops(coeff)];
>   for i from nops(coeff)-1 by -1 to 1 do
>     y := y * c + coeff[i];
>   od;
>   RETURN(y);
> end:
> input_list := [4, 3, 2, 1];
```
$$input\_list := [4, 3, 2, 1]$$

```
> Polynomial(5.0, input_list);
```
$$194.000$$

```
> Horner(5.0, input_list);
```
$$194.000$$

In order to test these procedures, we need a sample list of coefficients. The following command generates a random polynomial of degree 1000 in x.

```
> p2000 := randpoly(x,degree=2000,dense):
```

We have deliberately suppressed the output. Also, the algorithms expect a list of coefficients. This can be obtained from p as

```
>  q2000 := subs(x=1,convert(p2000,list)):
```

Now, using the Maple tools for measuring complexity, we determine which procedure runs relatively faster for a specific input.

```
>  st := time():
>  Horner(104567980000000.0, q2000 );
```
$$-.2348726858 \, 10^{27873}$$

```
>  time() - st;
```
$$.434$$

```
>  st := time():
>  Polynomial(104567980000000.0, q2000 );
```
$$-.2348726861 \, 10^{27873}$$

```
>  time() - st;
```
$$.833$$

Using Maple's computational complexity analysis tools, we can determine that the implementation of Horner's method of polynomial evaluation is marginally quicker than the implementation of the more traditional method of substitution, for the input covered here.

2.3   Number Theory

Maple offers an extensive library of functions and routines for exploring number theory. These facilities will help you to explore Sections 2.3, 2.4 and 2.5 of the text.

We begin our discussion of number theory by introducing modular arithmetic, greatest common divisors, and the extended Euclidean algorithm.

Basic Number Theory

To begin this subsection, we will see how to find the value of an integer modulo some other positive integer.

```
>  5 mod 3;
```
$$2$$

```
>  10375378 mod 124903;
```
$$8429$$

To solve equations involving modular congruences in one unknown, we can use the `msolve` function. For example, suppose we want to solve the problem: *What is the number that I need to multiply 3 by to get 1, modulo 7?* To solve this problem, we use the `msolve` function as follows.

```
>  msolve(3 * y = 1, 7);
```

$$\{y = 5\}$$

So, we find that $3 \cdot 5 = 15 = 14 + 1 \equiv 1 \pmod{7}$. Now, let us try to solve a similar problem, except that our modulus will be 6, instead of 7.

```
>  msolve(3 * y = 1, 6);
```

Now it appears that Maple has failed, but in fact, it has returned no solution, since no solution exists. In case there is any doubt, we will create a procedure to verify this finding.

```
>  CheckModSix := proc()
>    local i;
>    for i from 0 to 6 do
>       print(i, 3 * i mod 6);
>    od;
>  end:
>  CheckModSix();
```

$$0, 0$$
$$1, 3$$
$$2, 0$$
$$3, 3$$
$$4, 0$$
$$5, 3$$
$$6, 0$$

We note that $3y \equiv 0 \pmod 6$ or $3y \equiv 3 \pmod 6$, and hence $3y \equiv 1 \pmod 6$ will never have a solution. This can be attributed to the fact that $\gcd(3, 6) = 1$. As one final example of solving congruences, we shall construct a problem that has multiple solutions.

```
>  msolve(4 * x = 4, 10);
```

$$\{x = 6\}, \{x = 1\}$$

Greatest Common Divisors and Least Common Multiples

Maple provides a library routine `igcd` for computing the greatest common divisor of a set of integers. A few examples of the `igcd` function, along with other related functions, of Maple may be helpful.

```
>  igcd(3, 6);
```

```
> igcd(6, 4, 12);
```
$$2$$

Here, we compute the greatest common divisor of the integers from 10 to 100 inclusive.

```
> igcd(seq(i, i = 10..100));
```
$$1$$

There is a related function `ilcm` that computes the least common multiple. The following examples illustrate its use.

```
> ilcm(101, 13);
```
$$1313$$

```
> ilcm(6, 4, 12);
```
$$12$$

```
> ilcm(seq(i, i = 10..100));
```
$$69720375229712477164533808935312303556800$$

The last example calculates the least common multiple of the integers n in the range $10 \leq n \leq 100$.

Now to examine the relationships between least common multiples and greatest common divisors, we shall create a procedure called **IntegerRelations**.

```
> IntegerRelations := proc(a,b)
>    a*b, igcd(a,b), ilcm(a,b)
> end:
> IntegerRelations(6, 4);
```
$$24, 2, 12$$

```
> IntegerRelations(18, 12);
```
$$216, 6, 36$$

These examples illustrate the relationship

$$ab = \gcd\{a, b\} \cdot \operatorname{lcm}\{a, b\},$$

for non-negative integers a and b

The i in `igcd` and `ilcm` stands for "integer". The related functions `gcd` and `lcm` are more general and can be used to compute greatest common divisors and least common multiples of polynomials with rational coefficients. (They can also be used with integers, because an integer n can be identified with the polynomial $n \cdot x^0$.) The `igcd` and `ilcm` routines are optimized for use with integers, however, and may be faster for large calculations.

Now, having examined greatest common divisors, we may wish to address the problem of expressing a greatest common divisor of two integers as an integral combination of the integers. Specifically, given integers n and m, we may wish to express $\gcd(n, m)$ as a linear combination of m and n, such as $x \cdot n + y \cdot m$, where x and y are integers. To solve this problem, we will use the Extended Euclidean algorithm from Maple contained in the function `igcdex`, which stands for Integer Greatest Common Divisor using the Extended Euclidean algorithm. Since the Extended Euclidean Algorithm is meant to return three values, the `igcdex` procedure allows you to pass two parameters as arguments into which the result will be placed. By quoting them, we ensure we pass in their names, rather than any previously assigned value. You can access their values after calling `igcdex`. This is illustrated in the following example.

```
> igcdex(3,5, 'p', 'q');
```
$$1$$

```
> p; q;
```
$$2$$
$$-1$$

So, the desired linear combination is $2 \cdot 3 + (-1) \cdot 5 = 1 = \gcd(3, 5)$. We continue with two more examples.

```
> igcdex(2374, 268, 'x', 'y');
```
$$2$$

```
> x; y;
```
$$7$$
$$-62$$

```
> igcdex(1345, 276235, 'a', 'b');
```
$$5$$

```
> a; b;
```
$$5956$$
$$-29$$

Chinese Remainder Theorem

Maple can be used to solve systems of simultaneous linear congruences using the Chinese Remainder Theorem. (See Page 141 of the text.) To study problems involving the Chinese Remainder Theorem, and related problems, Maple offers the `chrem` function that computes the unique solution to the system

$$x \equiv a_1 \pmod{m_1}$$
$$x \equiv a_2 \pmod{m_2}$$
$$\cdots$$
$$x \equiv a_n \pmod{m_n}$$

of modular congruences. Specifically, we shall solve Example 5 (Page 140 of the text) using the Maple `chrem` function. Sun-Tzu's problem asks us to solve the following system of simultaneous linear congruences.

$$x \equiv 2 \pmod 3$$
$$x \equiv 3 \pmod 5$$
$$x \equiv 2 \pmod 7$$

The solution is easily computed in Maple, as follows.

```
> chrem([2, 3, 2], [3, 5, 7]);
                    23
```

The first list of variables in the `chrem` function contains the integers $a_1, a_2, \ldots, a_n$ and the second list of variables contains the moduli $m_1, m_2, \ldots, m_n$. The following, additional example illustrates the use of non-positive integers.

$$x \equiv 34 \pmod{98}$$
$$x \equiv -8 \pmod{23}$$
$$x \equiv 24 \pmod{47}$$
$$x \equiv 0 \pmod{39}$$

```
> chrem([34,-8,24,0],[98,23,47,39]);
                  826566
```

Having covered gcd's, modularity, the extended Euclidean algorithm and the Chinese Remainder Theorem, we move to the problem of factoring integers, which has direct practical applications to *cryptography*, the study of secret writing.

Factoring integers

To factor integers into their prime factors, the Maple number theory package `numtheory` must be loaded into memory, as follows.

```
>  with(numtheory):
```

If we wish to factor a number into its prime factors, we can use the Maple `ifactor` command. For example, we can factor 100 using `ifactor` as follows.

```
>  ifactor(100);
```

$$(2)^2\,(5)^2$$

```
>  ifactor(12345);
```

$$(3)\,(5)\,(823)$$

```
>  ifactor(1028487324871232341353586);
```

$$(2)\,(197)\,(72705239)\,(35903518878371)$$

By default, Maple uses the Morrison-Brillhart method, a factoring technique developed in the 1970's, to factor an integer into its prime factors. Beside using the Morrison-Brillhart method, Maple allows other methods of factorization to be used also. For instance, consider the following set of examples.

```
>  ifactor(1028487324871232, squfof);
```

$$(2)^6\,(19)\,(251)\,(3369703177)$$

```
>  ifactor(1028487324871232, pollard);
```

$$(2)^6\,(19)\,(251)\,(3369703177)$$

```
>  ifactor(1028487324871232, lenstra);
```

$$(2)^6\,(19)\,(251)\,(3369703177)$$

```
>  ifactor(1028487324871232, easy);
```

$$(2)^6\,(19)\,(251)\,(3369703177)$$

These examples illustrate several different methods of factorization available with Maple: the square-free method, Pollard's ρ method, and Lentra's elliptic curve method. The reader should explore these methods and various types of numbers that they factor efficiently or inefficiently. As an example, it is known that Pollard's method factors integers more efficiently if the factors are of the form $k \cdot m + 1$, where k is an optional third parameter to this method. It is left up to the reader to explore these alternative methods both using Maple and books on number theory.

The final factoring method which we will discuss, entitled **easy**, factors
the given number into factors which are easy to compute. The following
example illustrates this.

```
> ifactor(10284873248712323341353586);
```

$$(2)\,(197)\,(72705239)\,(35903518878371)$$

```
> ifactor(10284873248712323341353586, easy);
```

$$(2)\,(197)\,\_c22$$

The first method factors the given integer into complete prime factors,
where as the second method factors the number into small components
and returns _c22 indicating the other factor has 22 digits and is too hard
to factor. The time to factor is illustrated as follows.

```
> st:=time():
> ifactor(10284873247232341353586):
> time()-st;
```

$$.133$$

```
> st:=time():
> ifactor(10284873247232341353586, easy):
> time()-st;
```

$$.084$$

Primality Testing

Finding large primes is an important task in RSA cryptography, as we
shall see later. Here we shall introduce some of Maple's facilities for finding primes. We have already seen how to factor integers using Maple.
Although factoring an integer determines whether it is prime (since a
positive integer is prime if it is its only positive factor other than 1), factoring is not an efficient primality test. Factoring integers with 100 digits
is just barely practical today, using the best algorithms and networks of
computers, while factoring integers with 200 digits seems to be beyond
our present capabilities, requiring millions or billions of years of computer
time. (Here, we are talking about factoring integers not of special forms.
Check out Maple's capabilities. How large an integer can you factor in a
minute? In an hour? In a day?)

So, instead of factoring a number to determine whether it is a prime, we
use probabilistic primality tests. Maple has the `isprime` function which
is based upon such a test. When we use `isprime`, we give up the cer-
tainty that a number is prime if it passes these tests; instead, we know
that the probability this integer is prime is extremely high. Note that the

probabilistic primality test used by isprime is described in depth in Kenneth Rosen's textbook *Elementary Number Theory and its Applications* (3rd edition, published by Addison Wesley Publishing Company, Reading, Massachusetts, 1992).

We illustrate the use of the isprime function with the following examples.

```
>  isprime(101);
```
$$true$$

```
>  isprime(2342138342111);
```
$$true$$

```
>  isprime(23218093249834217);
```
$$false$$

Since this number is not too large for us to factor, we can use ifactor to check the result.

```
>  ifactor(23218093249834217);
```
$$(7)\,(3316870464262031)$$

The Maple procedure ithprime computes the ith prime number, beginning with the prime number 2.

```
>  ithprime(1);   # the first prime number
```
$$2$$

```
>  ithprime(2);   # the second prime number
```
$$3$$

```
>  ithprime(30000);
```
$$350377$$

The function ithprime produces prime numbers that are guaranteed to be prime. For small prime numbers, it simply looks up the result in an internal table, while for larger arguments, it operates recursively. This function should be used when an application needs to be certain of the primality of an integer and when speed is not an over-riding consideration.

In addition to these two procedures, Maple provides the nextprime and prevprime functions. As their names suggest, they may be used to locate prime numbers that follow or precede a given positive integer. For example, to find the first prime number larger than 1000, we can type

```
>  nextprime(1000);
```
$$1009$$

Similarly, the prime number before that one is

```
>  prevprime(");
```
$$997$$

Note that each of **nextprime** and **prevprime** is based on the function **isprime**, so their results are also determined probabilistically.

In general, to see the algorithm used by a procedure, set the **interface** parameter **verboseproc** equal to 2 and calling **eval** on the procedure. For example,

```
>  interface(verboseproc=2);
>  eval(nextprime);
```

proc(n)
    local i;
    option '*Copyright 1990 by the University of Waterloo*';
        if type(n, *integer*) **then**
            if $n < 2$ **then** 2
            else
                if irem($n, 2$) $= 0$ **then** $n + 1$ **else** $n + 2$ **fi**;
                for i **from** " **by** 2 **while** not isprime(i) **do od**;
                i
        fi
        elif type(n, *numeric*) **then** ERROR('*argument must be integer*')
        else 'nextprime(n)'
        fi
    end

These procedures provide several ways to generate sequences of prime numbers. A guaranteed sequence of primes can be generated quite simply using **seq**.

```
>  seq(ithprime(i), i=1..100);  # the first 100 primes
```
$2, 3, 5, 7, 11, 13, 17, 19, 23, 29, 31, 37, 41, 43, 47, 53, 59, 61, 67, 71, 73,$
$79, 83, 89, 97, 101, 103, 107, 109, 113, 127, 131, 137, 139, 149, 151,$
$157, 163, 167, 173, 179, 181, 191, 193, 197, 199, 211, 223, 227, 229,$
$233, 239, 241, 251, 257, 263, 269, 271, 277, 281, 283, 293, 307, 311,$
$313, 317, 331, 337, 347, 349, 353, 359, 367, 373, 379, 383, 389, 397,$
$401, 409, 419, 421, 431, 433, 439, 443, 449, 457, 461, 463, 467, 479,$
$487, 491, 499, 503, 509, 521, 523, 541$

The Euler ϕ-Function

The Euler ϕ function $\phi(n)$ counts the number of positive integers not exceeding n that are relatively prime go n. Note that since $\phi(n) = n - 1$

if, and only if, n is prime, we can determine whether n is prime by finding $\phi(n)$. However, this is not an efficient test.

In Maple, we can use the function phi in the numtheory package in the following manner.

```
>  phi(5);
```
$$4$$

```
>  phi(10);
```
$$4$$

```
>  phi(107);
```
$$106$$

This tells us that there are 4 numbers less than 5 that are relatively prime to 5, implying that 5 is a prime number. Since $\phi(5) = 4$ and $\phi(107) = 106$, we see that 5 and 107 are primes.

If we wished to determine all numbers $k_1, k_2, \ldots, k_m$ that have $\phi(k_i) = n$, we can use the invphi function of Maple. For example, to find all positive integers k such that $\phi(k) = 2$, we need only compute

```
>  invphi(2);
```
$$[3, 4, 6]$$

2.4   Applications of Number Theory

This section explores some applications of modular arithmetic and congruences. We discuss hashing, linear congruential random number generators, and classical cryptography.

Hash Functions

Among the most important applications of modular arithmetic is *hashing*. For an extensive treatment of hashing, the reader is invited to consult Volume 3 of D. Knuth's *The Art of Computer Programming*. Hashing is often used to improve the performance of search algorithms. This is important in many software systems such as in databases, and in computer languages translators (assemblers, compilers, and so on). Maple itself relies extensively, in its internal algorithms, upon hashing to optimize its performance.

Often, in a software system, it is necessary to maintain a so-called "symbol table". This is a table of fixed size in which various objects, or pointers to them, are stored. The number of objects input to the system is, in principle, unlimited, so it is necessary to map objects to locations in the symbol table in a many-to-one fashion. For this, a hashing function is used. Many types of hashing functions are used, but among the most effective are those based on modular arithmetic. Here, we'll look at how a hash function of the kind discussed in your textbook might be used in a simple minded way in the management of a simple symbol table. Our symbol table routines will do nothing more than install and search a table by means of a hash function.

The first thing to do is to decide on the size of the symbol table. We'll use Maple's `macro` facility to introduce a symbolic constant for this.

```
>   macro(hashsize = 101); # a prime number
```

Thus, our symbol table will have a fixed number `hashsize` of entries, not all of which need be occupied at a given time.

For this simple example, a "symbol" will simply be a string consisting exclusively of uppercase letters. For the symbol table itself we shall use a Maple array.

```
>   symtab := array(1..hashsize);
```
$$symtab := \mathrm{array}(1..101, [])$$

We'll define the hash function `Hash` to be used shortly, but first let's take a look at two procedures that will call `Hash`. The first is the function `Install`, used to enter a string into the symbol table.

```
>   Install := proc(s::string)
>     local hashval;
>     global symtab;
>
>     hashval := Hash(s);
>     symtab[hashval] := s;
>   end:
```

This procedure returns nothing; it is called only for the side effect of inserting the string argument into the symbol table. The second function is `Lookup`, used to search the symbol table for a string.

```
>   Lookup := proc(s::string)
>     local hashval, i;
>
>     hashval := Hash(s);
>     if symtab[hashval] = s then
>       RETURN(symtab[hashval]);
>     else
>       RETURN(NULL);
>     fi;
>   end:
```

The function Lookup computes the hash value of its argument, and returns the data stored at that address in the symbol table.

Now let's take a look at a hash function for strings that is based on modular arithmetic. We'll use a variant of the simple hash function discussed in the text. A very effective hash function for integers may be obtained by computing the remainder upon division by some modulus. Here, we'll use this idea by assigning to a string consisting of uppercase letters of the alphabet an integer, and then computing its value modulo the size of the symbol table. For this reason, we have chosen a symbol table size that is a prime number to maximize the "scattering" effect of the hash function, thus reducing the likelihood of collisions. (A major defect of our routines is the lack of any collision resolution strategy. You are asked in the exercises to repair this deficiency.) To compute an integer encoding of a string, we shall need the following procedure UpperToAscii that assigns to an uppercase character its ASCII value. First, we define a function UpperToNum that assigns to each uppercase character a number based on its position in the alphabet. This is not necessary here, but we'll reuse this function later on in this section.

```
>   alias( I = I );
>   alias( E = E );
>   alphabet := [A, B, C, D, E, F, G, H, I, J, K, L, M,
>                N, O, P, Q, R, S, T, U, V, W, X, Y, Z]:
>   for i from 1 to nops(alphabet) do
>     UpperToNum(op(i, alphabet)) := i - 1:
>   od:
```

Notice the special treatment given here to I and E. Each is a special symbol to Maple; E is used to denote the base of the natural logarithm, while I represents the imaginary unit. The alias calls above remove these special meanings. (In general, you should be very careful about how you redefine symbols having special meaning to Maple. We can do this here because we are certain that we do not need these two symbols to have their special meaning.) Here, now, is the ASCII conversion routine.

```
>   UpperToAscii := proc(s::string)
>     if not length(s) = 1 then
>       ERROR('argument must be a single character');
>     fi;
>     # The ASCII value of 'A' is 65.
>     RETURN(65 + UpperToNum(s));
>   end:
```

Here, finally, is the hash function.

```
>   Hash := proc(s::string)
>     local hashval, # return value
>           i;        # loop index
```

```
>
>    hashval := 0;
>    # Sum the ASCII values of the characters in s
>    for i from 1 to length(s) do
>      hashval := hashval + UpperToAscii(substring(s, i..i));
>    od;
>    # Compute the residue
>    hashval := hashval mod hashsize;
>    RETURN(hashval);
>  end:
```

We can see some of the hash values computed by our hashing function as follows.

```
>  Hash(MATH);
```

$$96$$

```
>  Hash(ALGEBRA);
```

$$90$$

```
>  Hash(FUNWITHMAPLE);
```

$$7$$

Now, a program might use the symbol table routines that we have developed here as follows. Given a list of strings to process in some way, the strings can be entered into the symbol table in a loop of some kind.

```
>  Input := ['BILL', 'HILLARY', 'CHELSEA', 'SOCKS', 'BILL'];
```

$$Input := [BILL, HILLARY, CHELSEA, SOCKS, BILL]$$

```
>  for s in Input do
>    if Lookup(s) = NULL then
>      Install(s);
>    fi;
>  od;
```

Here is what the symbol table looks like now that some entries have been installed.

```
>  eval(symtab);
```

$$\Big[?_1, ?_2, ?_3, ?_4, ?_5, ?_6, ?_7, ?_8, ?_9, ?_{10}, ?_{11}, ?_{12}, ?_{13}, ?_{14}, ?_{15}, ?_{16}, ?_{17}, ?_{18}, ?_{19},$$
$$?_{20}, ?_{21}, ?_{22}, ?_{23}, ?_{24}, ?_{25}, ?_{26}, ?_{27}, HILLARY, ?_{29}, ?_{30}, ?_{31}, ?_{32}, ?_{33}, ?_{34}, ?_{35},$$
$$?_{36}, ?_{37}, ?_{38}, ?_{39}, ?_{40}, ?_{41}, ?_{42}, ?_{43}, ?_{44}, ?_{45}, ?_{46}, ?_{47}, ?_{48}, ?_{49}, ?_{50}, ?_{51}, ?_{52},$$
$$?_{53}, ?_{54}, ?_{55}, ?_{56}, ?_{57}, ?_{58}, ?_{59}, ?_{60}, ?_{61}, ?_{62}, ?_{63}, ?_{64}, ?_{65}, ?_{66}, ?_{67}, ?_{68}, ?_{69},$$
$$?_{70}, ?_{71}, ?_{72}, ?_{73}, ?_{74}, ?_{75}, ?_{76}, ?_{77}, ?_{78}, ?_{79}, ?_{80}, ?_{81}, ?_{82}, ?_{83}, SOCKS, ?_{85},$$
$$?_{86}, ?_{87}, ?_{88}, BILL, ?_{90}, ?_{91}, ?_{92}, ?_{93}, ?_{94}, ?_{95}, ?_{96}, CHELSEA, ?_{98}, ?_{99}, ?_{100},$$
$$?_{101}\Big]$$

Each question mark (?) represents a cell in the table that is not yet occupied; its location is displayed as a subscript. The contents of occupied cells are shown here as strings.

To later extract strings from the symbol table for processing, or to determine whether a given string is already present in the table, the function Lookup is used.

```
>  Lookup('BILL');
```

$$BILL$$

```
>  Lookup('GEORGE');
```

Linear Congruential Pseudorandom Number Generators

Many applications require sequences of random numbers. They are important in cryptology and in generating data for computer simulations of various kinds. Often, random number streams are used as input to routines that generate random structures of different kinds, such as graphs or strings. It is impossible to produce a truly random stream of numbers using software only. (Software employs algorithms, and anything that can be generated by an algorithm is, by definition, not random.) Fortunately, for most applications, it is sufficient to generate a stream of **pseudorandom** numbers. This is a stream of numbers that, while not truly random, does nevertheless exhibit some of the same properties of a truly random number stream. Effective algorithms for generating pseudorandom numbers can be based on modular arithmetic. We examine here an implementation of a linear congruential pseudorandom number generator. This generates a sequence (x_n) of numbers x_n satisfying the system of equations

$$x_{n+1} = ax_n + c \pmod{m}$$

where a, c and the modulus m are some integer constants. Here, the first term x_1 of the sequence is initialized to some convenient value called the *seed*. One advantage the a pseudorandom number generator has in certain applications is that it can be reproduced simply by using the same seed. This is useful when the results are being used for test data that needs to be replicable from one instance of the test to the next.

To implement the generator, we start with a subroutine that does the modular arithmetic for us.

```
>  NextVal := proc(x,a,c,m)
>    RETURN((a * x + c) mod m);
>  end:
```

This simply computes the next value in the sequence, given the current value in the argument x. The generator itself is fairly simple. It simply initializes the data associated with the system, and constructs a loop to append the requested number of terms to the sequence returned. Our procedure LCPRNG takes two arguments, the length of the list to generate, and a *seed* or starting value used to initialize the generator.

```
> LCPRNG := proc(n::integer, seed::integer)
>    local prn_list, # list of pseudorandom numbers to return
>          modulus,
>          multiplier,
>          increment,
>          i,x;         # temporaries
>
>    # these could be globals instead,
>    # or passed as parameters
>    multiplier := 7^5;
>    modulus := 2^31 - 1;
>    increment := 66;
>
>    prn_list := NULL;
>    x := seed;
>    for i from 1 to n do
>      prn_list := prn_list, x;
>      x := NextVal(x, multiplier, increment, modulus);
>    od;
>    prn_list := [prn_list];
>    RETURN(prn_list);
> end:
```

To generate a list of 5 pseudorandom numbers, with seed 3, you can simply type:

```
> LCPRNG(5, 3);
```
$$[3, 50487, 848535075, 2037589511, 1992676381]$$

In practical use, you would likely choose the seed in a somewhat random fashion, say, based on the time of day. For instance, the following little routine produces an integer based on the CPU time of your Maple session.

```
> SeedIt := proc()
>    trunc(1000 * time());
> end;
```
$$SeedIt := \mathbf{proc}()\, \mathrm{trunc}(1000\, \mathrm{time}())\ \mathbf{end}$$

You can use it to generate somewhat unpredictable seed values for LCPRNG, as follows.

```
> LCPRNG(5, SeedIt());
```
$$[9116, 153212678, 212586459, 1675311518, 1302587275]$$

Classical Cryptography

We are going to examine here a way to implement an affine cipher in Maple. We'll need to convert between letters and numbers, as the ciphers we'll examine are based on arithmetic modulo 26. The UpperToNum function presented in the discussion of hashing in the previous section will serve well in one direction, but we shall also require its inverse NumToUpper.

```
>   for i from 0 to nops(alphabet) - 1 do
>     NumToUpper(i) := op(i + 1, alphabet):
>   od:
```

A general affine cipher has the form

$$f(p) = (ap + b) \pmod{26},$$

where the pair (a, b) is the *key* to the cipher. The argument p is the integer code for some plain text that is to be encrypted. For decryption to be feasible, the key must be chosen so that f is a bijection. This amounts to choosing a to be relatively prime to the modulus 26.

We shall use a helper function CryptChar to process a single character.

```
>   CryptChar := proc(s::string, key::[integer, integer])
>     local mult, # the multiplier
>           trans;# the translator
>
>     if not length(s) = 1 then
>       ERROR('argument must be a single character');
>     fi;
>     mult := key[1];
>     trans := key[2];
>     RETURN(NumToUpper((UpperToNum(s) * mult + trans) mod 26));
>   end:
```

This procedure encrypts single characters.

The cipher itself simply loops over all the character in the string input, and passes the individual calculations to Cryptchar.

```
>   AffineCypher := proc(s::string, key::[integer, integer])
>     local i,         # loop variable
>           multiplier,
>           translator,
>           ciphertext;# the encrypted text
>     ciphertext := NULL;
>     for i from 1 to length(s) do
>       ciphertext := cat(ciphertext,
>           CryptChar(substring(s, i..i), key));
>     od;
>     RETURN(ciphertext);
>   end:
```

Let's see how this works with a very regular string ABCDE and a few different keys:

```
>  AffineCypher(ABCDE, [1,3]); # Caesar cipher
```
$$DEFGH$$

```
>  AffineCypher(ABCDE, [3,0]);
```
$$ADGJM$$

```
>  AffineCypher(ABCDE, [3,3]);
```
$$DGJMP$$

Try this with various other keys. To encrypt the Maple string MATHISFUN, using the key $(3, 2)$, we can type

```
>  AffineCypher(MATHISFUN, [3,2]);
```
$$MCHXAERKP$$

An important observation is the the decryption function for an affine cipher is itself another affine cipher. Suppose that we are encrypting data with the key $(7, 2)$.

```
>  AffineCypher(ABC, [7,2]);
```
$$CJQ$$

To decipher the cipher text, we need to compute the inverse cipher. This is just the affine cipher with key $(15, 22)$, as later computations will show.

```
>  AffineCypher(CJQ, [15,22]);
```
$$ABC$$

Computing the key to the inverse cipher involves solving an affine congruence, modulo the alphabet size (here, 26). For our example, the encryption key was $(7, 2)$, corresponding to the congruence

$$y \equiv 7x + 2 \pmod{26}.$$

To compute the key for the decryption function, we need to solve this equation for x, modulo 26. We can use the solve procedure to do this in Maple, as follows.

```
>  x := 'x': y := 'y':
>  e := y = 7 * x + 2;
```
$$e := y = 7x + 2$$

```
>  solve(e, x);
```
$$\frac{1}{7}y - \frac{2}{7}$$

```
>   f := x = " mod 26;
```
$$f := x = 15\,y + 22$$

Thus, the defining equation for the inverse cipher is

$$y \equiv 15x + 22 \quad (\text{mod } 26),$$

to which corresponds the decryption key $(15, 22)$.

2.5 RSA Cryptography

We shall now show how to use Maple to implement the RSA cryptosystem. We shall use Maple to construct keys, and to encrypt and decrypt messages.

To construct keys in the RSA system, we need to find a pair of large primes, say, with 100 digits each. We shall explain how to do this later in this section. Since messages can be decrypted by anyone who can factor the product of these primes, the two primes must be large enough so that their product is extremely difficult to factor. A 200 digit integer fits the bill since factoring requires an extremely large amount of computer time.

Because the use of very large prime numbers would make our examples impractical *as examples*, we shall illustrate the RSA system using smaller primes, and then discuss, separately, how you can use Maple to generate large prime numbers.

Implementing the RSA system involves two steps.

1. Key generation

2. The encryption algorithm

Let us first consider key generation. The first step is to choose two distinct, large prime numbers, p and q, each of about 100 digits. From these, we must produce the public key, which consists of the public modulus $n = pq$, and the public exponent e, as well as the private key, consisting of the public modulus n, and the inverse of e modulo $\phi(n)$, where ϕ is Euler's ϕ-function. (The public modulus n is not really a part of the private key, but it does no harm to include it, and makes the implementation of the encryption engine below a little cleaner.) Since e is unrelated to the primes p and q, it can be generated in a number of ways. Two popular choices for e in real systems are 3 and the 4th Fermat number $F_4 = 2^{2^4} + 1 = 65537$. Another approach is to generate a random prime number that does not

divide $\phi(n)$. For our simple implementation below, we shall simply take e
to be the constant 13. Here is a Maple procedure to handle key generation
for us.

```
>  GenerateKeys := proc(p::integer, q::integer)
>     local n,   # public modulus
>           e,   # public exponent
>           d,   # d * e = 1 (mod phin)
>           phin; # phi(n) = (p - 1)(q - 1)
>
>     n := p * q; # Compute the public modulus
>     phin := (p - 1) * (q - 1);
>     e := 13; # This could be generated randomly
>
>     # Compute d such that e * d = 1 (mod phin)
>     d := op(1, op(Roots(e * x - 1) mod phin));
>
>     RETURN([[n, e], [n, d]]);
>  end:
```

This function returns both the public and private keys in a two member
list of the form [public_key, private_key]. Note that each entry is
itself a two member list of integers. In fact, it is useful to introduce a
Maple type for keys, both public and private. This allows us to write
clean code and still have procedures check their arguments.

```
>  'type/rsakey' := proc(obj)
>     type(obj, [posint, posint]);
>  end:
```

Thus, to generate keys using the prime numbers $p = 43$ and $q = 59$, we
can type

```
>  keys := GenerateKeys(43, 59);
```
$$keys := [[2537, 13], [2537, 937]]$$

and retrieve the public key and private key pairs, using op.

```
>  public_key := op(1, keys);
```
$$public\_key := [2537, 13]$$

```
>  private_key := op(2, keys);
```
$$private\_key := [2537, 937]$$

Our type definition above allows us to test the type of an object to determine whether it has the form of an RSA public or private key.

```
>  type(public_key, rsakey);
```
$$true$$

```
> type(private_key, rsakey);
```

true

In a practical RSA implementation, we would likely use some of the techniques discussed at the end of this section to incorporate into our GenerateKeys procedure the generation of the primes p and q as well, rather than passing them as arguments.

Now that we have seen how to generate RSA keys, we shall discuss an implementation of the encryption engine of the RSA scheme. The code is fairly simple.

```
> RSA := proc(key::rsakey, msg::list(posint))
>    local ct, # cipher text; returned
>          pe, # public exponent
>          pm, # public modulus
>          i;  # loop index
>
>    # Extract key information
>    pm := key[1];
>    pe := key[2];
>
>    ct := [];
>    for i from 1 to nops(msg) do
>      ct := [op(ct), msg[i]^pe mod pm];
>    od;
>
>    RETURN(ct);
> end:
```

The first argument to RSA is the key. The second argument is the message to be processed, in the form of a list of positive integers.

Let's look at how we can use RSA to transmit a message securely. First, a message in English or any other natural language must be encoded as a list of positive integers. To encode an English message, we can assign to each letter of the alphabet a two digit number, indicating the position of the letter in the alphabet. (So, for instance, $A \rightarrow 01$, $B \rightarrow 02$, and so on.) Then we break the resulting string of digits into blocks of four digits each. The list consisting of these four digit blocks will be the input to RSA. The block length must be chosen so that, after conversion, the largest integer produced is less than the modulus n. Here, we have $n = 2537$, and the largest block that can be produced is 2525 for "ZZ".

For a specific example, consider the message "STOP HERE". This is encoded as the list $[1819, 1415, 0805, 1705]$.

Now, to transmit the message "STOP HERE", the sender uses his or her private key `private_key`, computed above, to encrypt the message as follows.

```
> ciphertext := RSA(private_key, [1819, 1415, 0805, 1705]);
```
$$ciphertext := [2210, 2182, 126, 507]$$

The recipient receives the encrypted text `ciphertext`. To decrypt it, the sender's public key is used, as follows.

```
> RSA(public_key, ciphertext);
```
$$[1819, 1415, 805, 1705]$$

The resulting list is decoded as the original message "STOP HERE".

Generating Large Primes

If we were to use small prime numbers p and q such as those used in our earlier example, there would be no real security. Anyone could factor n, the product of these primes, and then easily find the decrypting key d from the encrypting key e. However, using Maple's prime generation and factoring facilities, we can generate fairly large prime numbers for use in an RSA public key. Remember that what is needed is a pair of prime numbers, each of about 100 digits. Moreover, they should be selected in an unpredictable fashion. To do this in Maple, we can use the **rand** procedure to produce a random 100 digit number, and then use the **nextprime** function to choose the smallest prime number that exceeds it. This will guarantee that the prime number has at least 100 digits.

For example, to choose two prime numbers of 100 digits each, we can proceed as follows.

```
> BigInt := rand(10^99..10^100): # get a random 100 digit number
> a := nextprime(BigInt());
```
$a := 40813211106932703436330736974742561435635584587189\backslash$
$\qquad 767467538305380320622220857229741217768604258122967$

```
> b := nextprime(BigInt());
```
$b := 44092598119526553100754871637971794904570391695941\backslash$
$\qquad 6008843057167496049883408581292045791645364177289$

Note that even generating such large prime numbers is not cheap.

```
> st := time():
> nextprime(BigInt()):
> time() - st;
```
$$10.083$$

It is left to the reader to incorporate these ideas in an improved version of our GenerateKeys procedure.

2.6 Base Expansions

To convert numbers from a representation in one base to a representation in another base, the Maple procedure convert can be used. For example, to convert the number 22 to its base 2 (binary) representation, we can type

```
>  convert(22, base, 2);
```

$$[0, 1, 1, 0, 1]$$

while the hexadecimal representation can be obtained by using the command

```
>  convert(22, base, 16);
```

$$[6, 1]$$

Conversion to a nonstandard base, such as 73 is also possible:

```
>  convert(1237765, base, 73);
```

$$[50, 19, 13, 3]$$

These examples illustrate that the output of the conversion is a list of elements ordered from lowest base value (the "ones" column) on the left, followed by the increasingly higher powers on the right. Converting integers to one of the "standard" bases from base 10 is fairly simple. The following examples illustrate some of these conversions.

```
>  convert(22, binary);
```

$$10110$$

```
>  convert(34, octal);
```

$$42$$

```
>  convert(1050, hex);
```

$$41A$$

Used in this form, the convert command provides more readable output, but it is limited to the commonly used bases shown here. Also, we can convert from a non-decimal base into base 10, using the "standard" bases of binary, octal, and hexadecimal.

```
>  convert(1001001,  decimal, binary);
```

$$73$$

```
>  convert('42E', decimal, hex);
```
$$1070$$

To convert from arbitrary bases into a decimal value, we shall construct a function that takes the list output that convert(int, base, n) returns.

```
>  MyConvert:=proc(L::list, base::integer)
>     local i, dec_value;
>     dec_value:=0;
>     for i from 1 to nops(L) do
>        dec_value := dec_value + L[i]*(base^(i-1));
>     od;
>  end:
>  t:=convert(145743, base, 73);
```
$$t := [35, 25, 27]$$

```
>  MyConvert(t, 73);
```
$$145743$$

The examples we have given illustrate that we may convert from any integer base into decimal, and from decimal to any integer base.

Maple can do arbitrary base conversions, at the cost of simplicity of representation (for bases larger than 26, we run out of 'digits"). For arbitrary base conversions, both input and output integers are represented as (Maple) lists of digits. For example, the base 10 integer 1996 is represented as the list [1, 9, 9, 6]. Each such digit must be a non-negative integer less than the input base, and each digit is interpreted as a base 10 integer. For instance, the base 10 number 263518331 can be written as a polynomial in 26 as

$$11 + 2 \cdot 26^2 + 17 \cdot 26^3 + 4 \cdot 26^4 + 22 \cdot 26^5$$

and is represented as the list [11, 0, 2, 17, 4, 22] of its "digits" in base 26. To convert this number to, say, base 99, we can type the following.

```
>  convert([11, 0, 2, 17, 4, 22], base, 26, 99);
```
$$[32, 87, 57, 73, 2]$$

The output should be interpreted as a list of digits for the base 99 number

$$32 + 87 \cdot 99^1 + 57 \cdot 99^2 + 73 \cdot 99^3 + 2 \cdot 99^4.$$

The same thing can be done with any pair of bases.

2.7   Matrices

Maple provides several matrix construction mechanisms and numerous operations to manipulate them. The simplest way to construct a matrix in Maple is by representing each row as a list and using the `matrix` command from the `linalg` package.

```
>   with(linalg):
>   a1:=matrix(
>     [[1, 2, 3, 4, 5], [6, 5, 4, 3, 2],
>      [2, 4, 6, 8, 0], [9, 7, 5, 3, 1]]
>   );
```

$$a1 := \begin{bmatrix} 1 & 2 & 3 & 4 & 5 \\ 6 & 5 & 4 & 3 & 2 \\ 2 & 4 & 6 & 8 & 0 \\ 9 & 7 & 5 & 3 & 1 \end{bmatrix}$$

Matrix expressions can be constructed as in

```
>   2*a1 + 3;
```

$$2\,a1 + 3$$

To complete the operations and view the results use the `evalm` function as in

```
>   evalm(2*a1+3);
```

$$\begin{bmatrix} 5 & 4 & 6 & 8 & 10 \\ 12 & 13 & 8 & 6 & 4 \\ 4 & 8 & 15 & 16 & 0 \\ 18 & 14 & 10 & 9 & 2 \end{bmatrix}$$

Maple also allows us to create matrices of common types, such as symmetric, antisymmetric, diagonal, identity or sparse.

```
>   a2:=array(1..5, 1..5, identity);
```

$$a2 := \mathrm{array}(identity, 1..5, 1..5, [])$$

```
> evalm(a2);
```

$$
\begin{bmatrix}
1 & 0 & 0 & 0 & 0 \\
0 & 1 & 0 & 0 & 0 \\
0 & 0 & 1 & 0 & 0 \\
0 & 0 & 0 & 1 & 0 \\
0 & 0 & 0 & 0 & 1
\end{bmatrix}
$$

```
> a3:=array(1..4, 1..4,
>    [(1,1)=4, (2, 2)=90, (3,3)=-34, (4,4)=103],
>    diagonal);
```

$$
a3 := \begin{bmatrix}
4 & 0 & 0 & 0 \\
0 & 90 & 0 & 0 \\
0 & 0 & -34 & 0 \\
0 & 0 & 0 & 103
\end{bmatrix}
$$

We may also exponentiate a square matrix any number of times, as shown by the following two examples.

```
> evalm(a3^2);
```

$$
\begin{bmatrix}
16 & 0 & 0 & 0 \\
0 & 8100 & 0 & 0 \\
0 & 0 & 1156 & 0 \\
0 & 0 & 0 & 10609
\end{bmatrix}
$$

```
> evalm(
>   matrix([[1,5,3], [-6, -4, 3], [0, 107, 4]])^5
> );
```

$$
\begin{bmatrix}
-523319 & -5503 & 231399 \\
-581982 & -1498792 & 304743 \\
980976 & 9888191 & -805328
\end{bmatrix}
$$

As a note of caution, since matrix multiplication is non-commutative, Maple has a special multiplication function designed specifically for matrices, whose name is &*. The usage is outlined in the following example.

```
> m1:=matrix([[1,86,3],[52, -3, 8], [-5, 34, 12]]);
```

$$
m1 := \begin{bmatrix}
1 & 86 & 3 \\
52 & -3 & 8 \\
-5 & 34 & 12
\end{bmatrix}
$$

```
> m2:=matrix([[4, 0, 6], [ 2, 7, -4], [18, 5, 13]]);
```

$$m2 := \begin{bmatrix} 4 & 0 & 6 \\ 2 & 7 & -4 \\ 18 & 5 & 13 \end{bmatrix}$$

```
>
> evalm(m1 &* m2);
```

$$\begin{bmatrix} 230 & 617 & -299 \\ 346 & 19 & 428 \\ 264 & 298 & -10 \end{bmatrix}$$

```
> evalm(m2 &* m1);
```

$$\begin{bmatrix} -26 & 548 & 84 \\ 386 & 15 & 14 \\ 213 & 1975 & 250 \end{bmatrix}$$

We now will examine how Maple can be used to solve a problem similar to that outlined in Example 6 on page 154 of the text. To solve this question, we first create a procedure that will keep track of the number of steps required to multiply two matrices.

```
> MyMatrixMult:=
>    proc(A::matrix, B::matrix, prev_steps::integer)
>    local i, j, k, q, C, steps;
>    steps:=prev_steps;
>    C:=matrix(rowdim(A), coldim(B), zeroes);
>    for i from 1 to rowdim(A) do
>    for j from 1 to coldim(B) do
>      C[i,j]:=0;
>      for q from 1 to coldim(A) do;
>        C[i,j]:=C[i,j]+A[i,q]*B[q,j];
>        steps:=steps+1;
>      od;
>    od;
>    od;
>    [evalm(C), steps];
>    end:

> MyMatrixMult(
>    matrix(4, 3, [1,0,4,2,1,1,3,1,0,0,2,2]),
>    matrix(3, 2, [2,4,1,1,3,0]),
>    0 );
```

$$\left[\begin{bmatrix} 14 & 4 \\ 8 & 9 \\ 7 & 13 \\ 8 & 2 \end{bmatrix}, 24 \right]$$

Having created this algorithm, we now will examine the best possible arrangement of matrix products so that computation is kept to a minimum. To make our sample size of matrices fairly large, we will use the `randmatrix` command, which produces matrices filled with random integers.

```
>   A1:=randmatrix(20,15):
>   A2:=randmatrix(15,35):
>   A3:=randmatrix(35,10):
>   MyMatrixMult(
>     A1,
>     MyMatrixMult(A2, A3, 0)[1],
>     MyMatrixMult(A2, A3, 0)[2])[2];
```
$$8250$$

```
>   MyMatrixMult(
>     MyMatrixMult(A1, A2, 0)[1],
>     A3,
>     MyMatrixMult(A1, A2, 0)[2])[2];
```
$$17500$$

Zero-One Matrices

Using Maple, we can create and manipulate zero-one matrices in a manner similar to integer valued matrices. In our exploration of zero-one matrices, we will create a zero-one matrix in a form that can be manipulated in Maple, then proceed to create the meet, join and Boolean product functions for zero-one matrices.

To create a Boolean matrix, we will define 1 to be `true`, and 0 to be `false`. This will allow Maple to apply boolean functions on each element in the matrix via the `bsimp` function of the logic package, as illustrated in the next example.

```
>   with(logic):
>   bsimp(true &and false);
```
false

```
>   bsimp(true &or false);
```
true

We now move on to constructing a boolean matrix. To do this, we will use the `matrix` function as was used before, entering the matrix as in zero-one form, then converting it to a Maple boolean form, using the `map` command.

```
>  with(linalg):
>  B1:=matrix(3,3,[1,0,1,1,1,0,0,1,0]);
```

$$B1 := \begin{bmatrix} 1 & 0 & 1 \\ 1 & 1 & 0 \\ 0 & 1 & 0 \end{bmatrix}$$

```
>  int_to_bool(1):=true:
>  int_to_bool(0):=false:
>  bool_to_int(true):=1:
>  bool_to_int(false):=0:
>  B2:=map(int_to_bool, B1);
```

$$B2 := \begin{bmatrix} true & false & true \\ true & true & false \\ false & true & false \end{bmatrix}$$

```
>  map(bool_to_int, B2);
```

$$\begin{bmatrix} 1 & 0 & 1 \\ 1 & 1 & 0 \\ 0 & 1 & 0 \end{bmatrix}$$

Having created a boolean matrix, both in zero-one format and Maple `true/false` format, we shall now create procedures for the boolean meet and join, as outlined on Page 157 of the textbook.

```
>  BoolMeet:=proc(A::matrix, B::matrix)
>    local i, j, C;
>    C:=matrix(rowdim(A), coldim(A), zeroes);
>    for i from 1 to rowdim(A) do
>      for j from 1 to coldim(A) do
>      C[i,j]:=bsimp(int_to_bool(A[i,j]) &and int_to_bool(B[i,j]));
>      od:
>    od;
>    map(bool_to_int,C);
>  end:
>  B3:=matrix(3, 3, [1, 0,0,1,1,1,0,0,0]);
```

$$B3 := \begin{bmatrix} 1 & 0 & 0 \\ 1 & 1 & 1 \\ 0 & 0 & 0 \end{bmatrix}$$

```
>  BoolMeet(B1, B3);
```

$$\begin{bmatrix} 1 & 0 & 0 \\ 1 & 1 & 0 \\ 0 & 0 & 0 \end{bmatrix}$$

```
>  BoolJoin:=proc(A::matrix, B::matrix)
>    local i, j, C;
>    C:=matrix(rowdim(A), coldim(A), zeroes);
>    for i from 1 to rowdim(A) do
>      for j from 1 to coldim(A) do
>      C[i,j]:=bsimp(int_to_bool(A[i,j]) &or int_to_bool(B[i,j]));
>      od;
>    od;
>    map(bool_to_int,C);
>  end:
>  BoolJoin(B1, B3);
```

$$\begin{bmatrix} 1 & 0 & 1 \\ 1 & 1 & 1 \\ 0 & 1 & 0 \end{bmatrix}$$

Having implemented the Boolean join and meet function, we conclude this subsection on zero-one matrices by implementing the Boolean product function, which is outlined on Page 158 of the text.

```
>  BoolProd:=proc(A::matrix, B::matrix)
>    local i, j, k, C;
>    C:=matrix(rowdim(A), coldim(B), zeroes);
>    for i from 1 to rowdim(A) do
>      for j from 1 to coldim(B) do
>        C[i,j]:=false;
>        for k from 1 to coldim(A) do
>          C[i,j]:=bsimp(
>            C[i,j]
>            &or (int_to_bool(A[i,k])
>                 &and int_to_bool(B[k,j])
>            )
>                        );
>      od;
>      od;
>    od;
>    map(bool_to_int, C);
>  end:
>  I1:=matrix(3, 2, [1,0,0,1,1,0]);
```

$$i1 := \begin{bmatrix} 1 & 0 \\ 0 & 1 \\ 1 & 0 \end{bmatrix}$$

```
>  I2:=matrix(2, 3, [1,1,0,0,1,1]);
```

$$i2 := \begin{bmatrix} 1 & 1 & 0 \\ 0 & 1 & 1 \end{bmatrix}$$

```
>  I3:=BoolProd(I1, I2);
```

$$i3 := \begin{bmatrix} 1 & 1 & 0 \\ 0 & 1 & 1 \\ 1 & 1 & 0 \end{bmatrix}$$

2.8 Computations and Explorations

For this section on the *Computations and Explorations* section of the textbook, we shall cover five representative questions; those being questions 1, 3, 4, 6, and 7, dealing with material on factorization and primality testing.

1. Determine whether $2^p - 1$ is prime for each of the primes p not exceeding 100.

Solution: To solve this problem, we'll write a Maple program that tests each prime less than or equal to a given to see whether $2^p - 1$ is prime, and produces a list of those primes for which it is. This is a good example of the use of the **nextprime** routine for looping over the list of primes up to some value.

```
>  Q1 := proc(n::integer)
>     local cur_prime, mlist;
>     cur_prime := 2;
>     mlist := NULL;
>     while cur_prime <= n do
>       if isprime(2^cur_prime - 1) then
>         mlist := mlist, cur_prime;
>       fi;
>       cur_prime := nextprime(cur_prime);
>     od;
>     mlist := [mlist];
>     RETURN(mlist);
>  end:
```

To check the primes not exceeding 100, we type the following.

```
>  Q1(100);
```

$$[2, 3, 5, 7, 13, 17, 19, 31, 61, 89]$$

For another approach, consider the `mersenne` procedure from the `numtheory` package; it is based on a table lookup algorithm. Using it, you can compute the 10th Mersenne prime, for example, with a call like.

```
>  numtheory[mersenne]([10]);
```
$$6189700196426690137449562111$$

Because it relies on a table in the Maple library, you cannot access very large Mersenne primes. See the help page for `numtheory,mersenne` for more information on this routine.

It is of note that there is a better test, called the *Lucas-Lehmer test*, that is more efficient and can be implemented in Maple. For a complete description of the algorithm, consult Rosen's text on Number Theory (as referenced earlier in the chapter).

3. Find as many primes of the form $n^2 + 1$ where n is a positive integer as you can. It is not know whether there are infinitely many such primes.

Solution: Looking at this question, we again construct a Maple procedure to aid in our search of primes of this form.

```
>  Q3 := proc(n)
>    local i;
>    for i from 1 to n do
>    if isprime(i^2 + 1) then
>      printf('For i = %d, the number i^2 + 1 = %d is prime\n',
>           i, i^2 + 1);
>    fi;
>    od;
>  end:
```

To save space, we'll only do a small calculation.

```
>  Q3(10);
```

```
For i = 1, the number i^2 + 1 = 2 is prime
For i = 2, the number i^2 + 1 = 5 is prime
For i = 4, the number i^2 + 1 = 17 is prime
For i = 6, the number i^2 + 1 = 37 is prime
For i = 10, the number i^2 + 1 = 101 is prime
```

You can try `Q3(1000)` or `Q3(10000)`, which yield several pages of output. What can you observe about the output?

4. Find 10 different primes, each with 100 digits.

Solution: Maple's `nextprime` function makes this too easy, so we shall construct 10 *random* primes of 100 digits each. It is still not too hard, though; the only extra wrinkle is that we must guard against including a prime in our list more than once. Our procedure here is a little more general, since it is almost trivial to have it take, as an argument, the number of primes to produce.

```
> PrettyBigPrimes := proc(howmany::integer)
>    local ptab,  # table of 10 primes
>          BigInt,# random 10 digit number generator
>          n,     # loop variable
>          p;     # index into ptab
>    BigInt := rand(10^99..10^100);
>    ptab := {};
>    for n from 1 to howmany do
>      p := nextprime(BigInt());
>      # loop until we get a new one
>      while member(p, ptab) do
>         p := nextprime(p);
>      od;
>      ptab := ptab union {p};
>    od;
>    # convert to a list and return it
>    RETURN(convert(ptab, 'list'));
> end:
```

To save space, we'll show the output of calculating only 3 primes.

```
> PrettyBigPrimes(3);
```

$$[93079206249473499510535300861464863071981555907634664\backslash$$
$$29392673709525428510973272600608981219688541373, 44092\backslash$$
$$59811952655310075487163797179490457039169594160088430\backslash$$
$$57167496049883408581292045791645366417 7289, 4081321110\backslash$$
$$69327034363307369747425614356355845871897674675383053\backslash$$
$$803206222208572297412176860425812 2967]$$

However, to illustrate that this is a fairly expensive computation, we show a timed run for a calculation of 10 primes.

```
> st := time():
> PrettyBigPrimes(10):
> time() - st;
```

$$23.366$$

6. Find a prime factor of each of 10 different 20-digit odd integers, selected at random. Keep track of how long it takes to find a factor of each of these integers. Do the same thing for 10 different 30-digit odd integers, 10 different 40-digit odd integers, and so on, continuing as long as possible.

Solution: For this question, we first need to use Maple to select random numbers in a specific size range; that is, random 20-digit numbers (and 30-digit and 40-digit and so on). To do this, we will use the `rand` function of Maple.

```
> Twenty:=rand(10^19..10^20):
> Twenty();
```
$$25905687273574097027$$

The next step in solving Question 6 is to create some pseudocode that may help us in determining the desired results:

(a) Create a random number

(b) Verify the number is odd. If the random number is even, simply add 1 to it to make it odd.

(c) Determine a factor of the random number (we need not find all prime factors).

(d) Record the time required, and return to step (a)

We have already outlined how to do step (a), and step (b) can be fulfilled by the use of the `mod` function. We can accomplish step (c) by the use of the `ifactor` function, and step (d) can be done using the Maple `time` function. We now move on to defining the procedure that is required.

```
> RandFactor := proc(num_digits::integer)
>    local i, temp, temp_fact,
>    total_time, Generator, st;
>    total_time := 0;
>    Generator := rand(10^(num_digits-1)..10^num_digits);
>    for i from 1 to 10 do
>      temp := Generator();
>      if (temp mod 2 = 0) then
>        temp := temp+1;
>      fi;
>      st := time();
>      temp_fact := ifactor(temp, easy);
>      total_time := total_time + time() - st;
>      print( temp = temp_fact );
>    od;
>    printf('\nTime: %a\n', total_time);
>  end:
> RandFactor(20);
```

$$34381304659501242715 = (5)\,(6876260931900248543)$$
$$44857856234450259381 = (3)\,(90670992487)\,(56807)\,(2903)$$
$$66273485070460234409 = (41)\,(39350981)\,(178397)\,(230257)$$
$$47229320612320189545 = (3)\,(5)\,(73)\,\_c17$$
$$26898380788553860081 = (82274348881)\,(32573)\,(10037)$$
$$66682595985479914973 = (13)\,(43)\,(241)\,(14155117)\,(34967951)$$
$$27140299651198097705 = (5)\,(11)\,(29)\,(1781579085989)\,(9551)$$

$$5930557905 9712269761 = (986283113)\,(295033)\,(203809)$$
$$35483826208735328545 = (5)\,(11)\,(85440431)\,(12119)\,(623071)$$
$$39794196134912187317 = (67)\,(257)\,(2311063135775143)$$

Time: 1.449

The reader can continue this exploration with 30-digit numbers and upwards, to determine the size of input that makes Maple factoring impractical.

7. Find all pseudoprimes to the base 2, that is, composite integers n such that $2^{n-1} \equiv 1 \pmod{n}$, where n does not exceed 10000?

Solution: Using Maple, this problem is relatively straightforward to solve. We can use a for-loop structure to increment our n variable, the isprime function to determine which values of n are composite, and the mod function to calculate the remainder of 2^{n-1} modulo n.

The Maple code is shown below. Because of the length of the output, we show here only the first 100 values. You can check values up to 1000, or even higher, yourself.

```
>   n := 'n':
>   for n from 1 to 100 do
>     if 2^(n-1) mod n = 1 and isprime(n) then
>       print(n);
>     fi;
>   od:
```

2.9   Exercises/Projects

Exercise 1. Test which is faster for computing the greatest common divisor of a collection of integers, the igcd or gcd function.

Exercise 2. How would you use Maple to generate the list of the first 100 prime numbers larger than one million?

Exercise 3. Investigate further the comparative performance of Horner's method (discussed in Section 2.2) and the method of substitution for polynomial evaluation.

Exercise 4. Use Maple to find the one's complement of an arbitrary integer. (See Page 135 of the text.)

Exercise 5. For which odd prime moduli is -1 a square? That is, for which prime numbers p does there exist an integer x for which $x^2 \equiv -1$ (mod p)?

Exercise 6. Use Maple to determine which numbers are perfect squares modulo n for various values of the modulus n. What conjectures can you make about the number of different square roots an integer has module n? (Hint: Use the Maple function `msqr`).

Exercise 7. Use Maple to find the base 2 expansion of the 4th Fermat number $F_4 = 2^{2^4} + 1$. Do the following for several large integers n. Compute the time required to calculate the remainder modulo n, of various bases b raised to the power F_4 (that is, to calculate b^{F_4} (mod n)) using two different methods: First, do the calculation by a straightforward exponentiation; then do it using the binary expansion of F_4 with repeated squarings and multiplications. Why do you think it is a good choice for the public exponent in the RSA encryption scheme?

Exercise 8. Modify the procedure `GenerateKeys` that we developed to produce the keys for the RSA system to incorporate the techniques for generating large random prime numbers. Make your procedure take as an argument a "security" parameter, as measured by the number of digits in the public modulus.

Exercise 9. Write Maple routines to encode and decode English sentences into lists of integers, as described in the section on RSA encryption. (You may ignore spaces and punctuation, and assume that all letters are uppercase.)

Exercise 10. Modify the symbol table management routines presented in this chapter to employ a collision strategy. Instead of storing the string data itself in the symbol table, use a list, and search the list linearly for a given input, after computing the location in the table by hashing. A reference for information on collision resolution strategies is Section 4.4 in K. Rosen, *Elementary Number Theory and its Applications, 3rd ed.*, Addison-Wesley, Reading, Massachusetts, 1992.

Exercise 11. (Class Project) The *Data Encryption Standard* (DES) specifies a widely used algorithm for private key cryptography. (You can find a description of this algorithm in Cryptography, Theory and Practice, by Douglas Stinson, published by CRC Press). Implement the DES in Maple.

Exercise 12. There are infinitely many primes of the form $4n + 1$ and infinitely many of the form $4N + 3$. Use Maple to determine for various values of x whether there are more primes of the form $4n + 1$ less than x than there are primes of the form $4n + 3$, or vice versa. What conjectures can you make from this evidence?

Exercise 13. Develop a procedure for determining whether Mersenne numbers are prime using the Lucas-Lehmer test as described in number theory books such as K. Rosen, *Elementary Number Theory and its Applications, 3rd ed.*, Addison-Wesley, Reading, Massachusetts, 1992. How many Mersenne numbers can you test for primality using Maple?

Exercise 14. Repunits are integers with decimal expansions consisting entirely of 1s (e.g. $111, 111111$, and 11111111111). Use Maple to factor repunits. How many prime repunits can you find? Explore the same question for repunits in different base expansions.

Exercise 15. Compute the sequence of pseudorandom numbers generated by the linear congruential generator $x_{n+1} = (ax_n + c) \pmod{m}$ for various values of the multiplier a, the increment c, and the modulus m. For which values do you get a period of length m for the sequence you generate? Can you formulate a conjecture about this?

Exercise 16. The Maple function tau (in the numtheory package) implements the function τ defined, for positive integers n by declaring that $\tau(n)$ is the number of positive divisors of n.

```
>  numtheory[tau](20);
```

$$6$$

Use Maple to study the function τ. What conjectures can you make about τ? For example, when is $\tau(n)$ odd? Is there a formula for $\tau(n)$? For which integers m does the equation $\tau(n) = m$ have a solution, for some integer n? Is there a formula for $\tau(mn)$ in terms of $\tau(m)$ and $\tau(n)$?

Exercise 17. Develop a procedure that solves Josephus' problem. This problem asks for the permutation describing the order in which soldiers are killed when n soldiers arrange themselves around a circle and repeatedly execute every mth soldier, given the integers m and n.

3 Mathematical Reasoning

In this chapter we describe how Maple can be used to help in the construction and understanding of mathematical proofs. Computational capabilities may seem not particularly relevant to the study of proofs. However, in reality, these capabilities can be helpful with proofs in many ways. In this chapter we describe how Maple can used to work with formal rules of inference. We describe how Maple can be used to help gain insight into constructive and non-constructive proofs. Moreover, we show how to use Maple to help develop proofs using mathematical induction, even demonstrating how Maple can be used both for the basis step and the inductive step in the proof of summation formulae. Moreover, we show how Maple can be used to compute terms of recursively defined sequences. We will also compare the efficiency of generating terms of such a sequence via inductive versus recursive techniques.

3.1 Methods of Proof

Although Maple cannot take theorems and output proofs for those theorems, it can take logical expressions and simplify them or determine characteristics such as whether a boolean expression is satisfiable or is a tautology. To work with logical expressions in Maple, we shall need to use some of the facilities provided by the `logic` package, a topic discussed more fully in Chapter 9.

Firstly, we will examine the logical operators of "and", "or", "not" and "implies". There is no system operator "implies"; to study conditionals, we must work with the inert boolean operators provided by the `logic` package. These all begin with the character & so, for instance, we use &and instead of and, and &not in place of not. See Chapter 9 for a more complete discussion of the distinction between these two sets of operators. Here are some examples of the use of the inert boolean operators.

```
>   with(logic):
>   a := &not(x1 &and x2);
```
$$a := \&\mathrm{not}\,(\mathit{x1}\,\&\mathrm{and}\,\mathit{x2})$$

```
>   b := (x3 &or x4) &implies (x5 &and x6);
```
$$b := (\mathit{x3}\,\&\mathrm{or}\,\mathit{x4})\,\&\mathrm{implies}\,(\mathit{x5}\,\&\mathrm{and}\,\mathit{x6})$$

We concern ourselves now with determining how Maple simplifies boolean expressions if we have them in combination. We begin with a simple "double negative" example.

```
>   c := &not (&not x7);
```
$$c := \&\text{not}(\&\text{not}(x7))$$

This can be simplified by the use of the `bsimp` function of Maple.

```
>   bsimp(c);
```
$$x7$$

Now, we move to a more complex example, in which the reader can confirm the correctness of the simplification, constructing a truth table.

```
>   d := &not (&not x8 &and &not x9);
```
$$d := \&\text{not}(\&\text{not}(x8) \,\&\text{and}\, \&\text{not}(x9))$$

```
>   bsimp(d);
```
$$x9 \,\&\text{or}\, x8$$

The next example illustrates the simplification of Modus Ponens. We first state that "p implies q" and "p is true".

```
>   e := (p &implies q) &and (p);
```
$$e := (p \,\&\text{implies}\, q) \,\&\text{and}\, p$$

Then we simplify the boolean expression,

```
>   bsimp(e);
```
$$p \,\&\text{and}\, q$$

determining that "q and p is true", and since we already knew "p was true", we have concluded that "q is true".

The `bsimp` function is a general simplifier for boolean expressions constructed using the inert boolean operators. It computes a simplified boolean expression equivalent to its argument. Consult Chapter 9 for more details on `bsimp`.

We can also use Maple to determine if an expression is a tautology, by using the `tautology` function in the "logic" package.

```
>   tautology(x1 &and x2);
```
$$\textit{false}$$

```
>   tautology((&not x3) &or x3);
```
$$\textit{true}$$

We now show how Maple can be used to gain more insight into some constructive proofs. Specifically, we will examine how Maple can be used to explore the conclusions of Example 20 (page 179) in the text, which is exploring how to construct a list of sequential composite numbers. We shall create the constructive algorithm outlined in the text, in order to explicitly generate this list of composite number.

```
>  MakeComposite := proc(n::integer)
>    local x,i, L;
>    L := {};
>    x := (n + 1)! + 1:
>    for i from 1 to n do
>        L := L union {(x + i)};
>    od;
>    L;
>  end:
>  MakeComposite(5);
```

$$\{723, 724, 725, 726, 722\}$$

```
>  MakeComposite(11);
```

$\{479001603, 479001604, 479001605, 479001606, 479001607,$
$479001608, 479001609, 479001610, 479001611, 479001612, 479001602$
$\}$

While Maple can be used to generate the list of n consecutive composite integers generated by the proof, it is not possible to use Maple to derive the proof itself. It should be noted that this argument does not provide the smallest set of n consecutive composite integers. However, given a positive integer n, you could use Maple to find the smallest sequence of n consecutive composite integers. (See Problem 3 in the Computations and Explorations section of the text, and the exercises at the end of this chapter.)

Now, we turn our attention to Example 21, which is the non-constructive existence proof of the fact that there are an infinite number of prime numbers. Now, since this proof is non-constructive, we cannot simply create an algorithm to generate a "larger prime" assuming the existence of a "largest prime". However, the key idea in the proof is to consider the primality of the integer $n! + 1$, and Maple can be used to pursue this a little further. Of course, it is possible that $n! + 1$ is itself a prime number but, even if it is not, its smallest prime factor must be larger than n. We can use Maple to find this smallest prime factor by factoring $n! + 1$ directly, using the Maple library routine ifactor. Let's examine a few of the numbers of this form.

```
>  for n from 1 to 10 do
>     ifactor(n! + 1);
>  od;
```

$$(2)$$
$$(3)$$
$$(7)$$
$$(5)^2$$
$$(11)^2$$
$$(7)\,(103)$$
$$(71)^2$$
$$(61)\,(661)$$
$$(19)\,(71)\,(269)$$
$$(11)\,(329891)$$

We can see from the output that, while some of these numbers are themselves prime, others are not, and from this, we can read off the smallest of their prime factors.

To determine the least prime factor of each of these integers, we can write a routine as follows.

```
>  with(numtheory): # define 'factorset'
>  LeastFactor := proc(n::integer)
>     min(op(factorset(n)));
>  end:
```

This uses the procedure `factorset`, from the `numtheory` package, to compute the set of factors of the integer input, and then simply selects its least member.

```
>  for n from 1 to 10 do
>     LeastFactor(n! + 1);
>  od;
```

$$2$$
$$3$$
$$7$$
$$5$$
$$11$$
$$7$$
$$71$$
$$61$$
$$19$$
$$11$$

Now, we confront our final example of using Maple for exploring mathematical theorems. In this case, we will examine Goldbach's conjecture: that is, "Every even integer greater than 4 can be expressed as a sum of two primes."

```
>  Goldbach := proc(p::integer)
>    local i,j,finished, next_i_value;
>    finished := false;
>    i := 0; j := 0;
>    while not finished do
>      next_i_value := false;
>      i := i+1;  j := i;
>      while not next_i_value do
>        if ithprime(i) + ithprime(j) = p then
>          printf('%d can be expressed as %d + %d\n',
>                   p, ithprime(i), ithprime(j));
>          finished := true;
>          next_i_value := true;
>        fi;
>        j := j+1;
>        if ithprime(j) >= p then
>          next_i_value := true
>        fi;
>      od;
>    od;
>  end:
>  Goldbach(12);

12 can be expressed as 5 + 7

>  Goldbach(24);

24 can be expressed as 5 + 19
```

Now, we create a procedure to examine the Goldbach conjecture more automatically.

```
>  ManyGoldbach := proc(startval::integer,finalval::integer)
>    local i;
>    for i from max(2, startval) to finalval do
>      Goldbach(2 * i);
>    od;
>  end:
>  ManyGoldbach(2,4);

4 can be expressed as 2 + 2
6 can be expressed as 3 + 3
8 can be expressed as 3 + 5

>  ManyGoldbach(20, 26);

40 can be expressed as 3 + 37
42 can be expressed as 5 + 37
44 can be expressed as 3 + 41
```

```
46 can be expressed as 3 + 43
48 can be expressed as 5 + 43
50 can be expressed as 3 + 47
52 can be expressed as 5 + 47
```

true

3.2   Mathematical Induction

It is possible to use Maple to assist in working out proofs of various mathematical assertions using mathematical induction. In fact, with Maple as your assistant, you can carry out the entire process of discovery and verification interactively. We'll demonstrate this here, first with a very simple example, to highlight the steps involved; then we'll examine a somewhat less trivial problem.

It is likely that one among the first examples of the use of mathematical induction that you encountered is the verification of the formula

$$1 + 2 + 3 + \cdots + n = \frac{n(n+1)}{2}$$

for the sum of the first n positive integers. Maple is ideally suited to proving formulae, such as this one, because the steps in an inductive proof involve symbolic manipulation. It is hardly necessary in this simple example, but you can use Maple to generate a large body of numerical data to examine.

```
>  seq(sum(i, i = 1..n), n = 1..30);
```
   $1, 3, 6, 10, 15, 21, 28, 36, 45, 55, 66, 78, 91, 105, 120, 136, 153, 171, 190,$
      $210, 231, 253, 276, 300, 325, 351, 378, 406, 435, 465$

By generating a sufficiently large set of numerical data, and with a little insight, you should eventually be able to guess the formula above. The output shows that n^2 is a little less than twice the sum of the first n integers for the values of n tested. From the pattern we see, we might guess that the correct formula is a quadratic function of n, solve for the coefficients, and then test whether this procedure produces the correct formula.

A useful technique for experimenting with such guesses is to generate lists of pairs consisting of the sequence you are interested in and various "guesses" that you come up with. To investigate the hypothesis that the formula is quadratic, you might start by generating a list of pairs such as the one that follows.

```
>   s := 's':
>   s := proc(n::integer)
>    local i;
>    sum(i, i = 1..n);
>   end:
>   seq([s(n), n^2], n=1..20);
```
$$[1, 1], [3, 4], [6, 9], [10, 16], [15, 25], [21, 36], [28, 49], [36, 64], [45, 81],$$
$$[55, 100], [66, 121], [78, 144], [91, 169], [105, 196], [120, 225],$$
$$[136, 256], [153, 289], [171, 324], [190, 361], [210, 400]$$

To explore whether the sum is a quadratic function of n, we can enter a generic quadratic in n, and solve for the coefficients a, b and c.

```
>   n := 'n'; # remove any value
```
$$n := n$$

```
>   q(n) := a * n^2 + b * n + c;
```
$$q(n) := a\,n^2 + b\,n + c$$

We need three equations to solve for three coefficients.

```
>   eqns := {seq(subs(n = k, q(n)) = s(k), k = 1..3)};
```
$$eqns := \{a + b + c = 1, 4\,a + 2\,b + c = 3, 9\,a + 3\,b + c = 6\}$$

Now we instruct Maple to solve these equations for the three coefficients.

```
>   solve(eqns, {a, b, c});
```
$$\{c = 0, b = \frac{1}{2}, a = \frac{1}{2}\}$$

Our original formula then becomes

```
>   subs(", q(n));
```
$$\frac{1}{2}\,n^2 + \frac{1}{2}\,n$$

At this point, you can use Maples ability to manipulate expressions *symbolically* to help to construct an inductive proof. Here is how an interactive proof of the formula above, by mathematical induction, can be carried out in Maple.

The general term of the sum is

```
>   genterm := n;
```
$$genterm := n$$

while the right hand side of the formula is

```
>   formula := n * (n + 1)/2;
```
$$formula := \frac{1}{2}\,n\,(n + 1)$$

We can use the subs procedure to check the basis step for the induction; here the base case is that in which $n = 1$.

```
>  subs(n = 1, genterm);
```
$$1$$

```
>  subs(n = 1, formula);
```
$$1$$

The results agree, so the basis step is established.

For the inductive step, we suppose the formula to be valid for $n = k$.

```
>  indhyp := subs(n = k, formula);
```
$$indhyp := \frac{1}{2} k (k + 1)$$

To sum $k + 1$ terms, we compute

```
>  indhyp + subs(n = k + 1, genterm);
```
$$\frac{1}{2} k (k + 1) + k + 1$$

Finally, the formula for $n = k + 1$ is

```
>  subs(n = k + 1, formula);
```
$$\frac{1}{2} (k + 1) (k + 2)$$

The results agree, so the inductive step is verified. The formula now follows by mathematical induction. Thus, you can see that, while Maple is not (yet) able to construct proofs entirely on its own, it is a very effective tool to use in an interactive proof construction.

Now let's consider a more complicated example. A formula for the sum

$$S = 1 \cdot 1! + 2 \cdot 2! + 3 \cdot 3! + \cdots + n \cdot n!$$

is much less obvious than in the preceding example. To discover one, we begin by generating some numerical data.

```
>  seq(sum(i * i!, i = 1..n), n = 1..20);
```
$$1, 5, 23, 119, 719, 5039, 40319, 362879, 3628799, 39916799, 479001599,$$
$$6227020799, 87178291199, 1307674367999, 20922789887999,$$
$$355687428095999, 6402373705727999, 121645100408831999,$$
$$2432902008176639999, 51090942171709439999$$

If a pattern is not immediately obvious, we can assist our intuition be generating a parallel sequence.

```
>   seq([n!, sum(i * i!, i = 1..n)], n = 1..10);
```

$$[1, 1], [2, 5], [6, 23], [24, 119], [120, 719], [720, 5039], [5040, 40319],$$
$$[40320, 362879], [362880, 3628799], [3628800, 39916799]$$

Staring at this a little renders obvious the fact that we are on to something. Let's just adjust this a little.

```
>   seq([(n + 1)!, sum(i * i!, i = 1..n)], n = 1..10);
```

$$[2, 1], [6, 5], [24, 23], [120, 119], [720, 719], [5040, 5039],$$
$$[40320, 40319], [362880, 362879], [3628800, 3628799],$$
$$[39916800, 39916799]$$

From this evidence, we should probably infer the conjecture that a formula for our sum is

$$S = (n + 1)! - 1.$$

The inductive proof can be carried out much as we did in the first example.

```
>   n := 'n': k := 'k': # clear values
>   S := (n + 1)! -1: genterm := n * n!:
```

The basis step is

```
>   subs(n = 1, genterm);
```

$$1$$

```
>   subs(n = 1, S);
```

$$1$$

The inductive step is

```
>   indhyp := subs(n = k, S); # induction step
```

$$indhyp := (k + 1)! - 1$$

```
>   indhyp + subs(n = k + 1, genterm);
```

$$(k + 1)! - 1 + (k + 1)(k + 1)!$$

```
>   subs(n = k + 1, S);
```

$$(k + 2)! - 1$$

Using a little algebraic manipulation, we see that the last two formulae are equal. This completes a proof *via* mathematical induction. We conclude that our guess at the formula is correct.

3.3   **Recursive and Iterative Definitions**

Maple functions can be defined in both procedurally (using the `proc` function) and explicitly (using the `->` notation), and each of these methods can involve recursive and iterative means of definitions. We begin our study using the `->` function of Maple. If we wished to define the polynomial function $a(n) = 3n^3 + 41n^2 - 3n + 101$, we would issue the following Maple command:

```
>   a:=n->3*n^3+41*n^2-3*n+101;
```
$$a := n \to 3\,n^3 + 41\,n^2 - 3\,n + 101$$

```
>   a(5);
```
$$1486$$

```
>   a(523);
```
$$440380222$$

Now, if we wished to define a function recursively, say $b(n) = b(n-1)^2 + 2b(n-1) + 6$, with the initial condition $b(0) = 2$, then we would enter

```
>   b:=n->b(n-1)^2+2*b(n-1)+6;
```
$$b := n \to \mathrm{b}(n-1)^2 + 2\,\mathrm{b}(n-1) + 6$$

```
>   b(0):=2;
```
$$\mathrm{b}(0) := 2$$

```
>   b(1);
```
$$14$$

If we wished to see a sequence of values for the function b, we can use the `seq` function of Maple to display output for a given range of input.

```
>   seq(b(i),i=1..7);
```
$$14, 230, 53366, 2848036694, 8111313016066523030,$$
$$65793398844610194510432568305653426966, 43287713315259\backslash$$
$$540779309909083428204826613190138309545789565399286 71\backslash$$
$$202819094$$

Now, we shall create a similar function to b, called `f1`, that will find Fibonacci numbers.

```
>   f1 := n->f1(n-1)+f1(n-2);
```
$$f1 := n \to \mathrm{f1}(n-1) + \mathrm{f1}(n-2)$$

```
>  f1(1) := 1;
```
$$f1(1) := 1$$

```
>  f1(2) := 1;
```
$$f1(2) := 1$$

```
>  f1(5);
```
$$5$$

```
>  seq(f1(i), i = 1..15);
```
$$1, 1, 2, 3, 5, 8, 13, 21, 34, 55, 89, 144, 233, 377, 610$$

While the -> notation for functions is convenient and intuitive, it does not offer all of the facilities for improving efficiency that are available using the proc command. To force Maple to calculate these values more efficiently, we use the remember option to procedure definitions effected using proc. This option requires Maple to "remember" any values for the procedure that it has already computed by storing them in a table.

```
>  f2 := proc(n::integer) option remember;
>    if n <= 2 then RETURN( 1 ) fi;
>    f2(n-1) + f2(n-2);
>  end:
```

So, this procedural method encompasses both the base cases (when $n \leq 2$) and the inductive cases (as in the else condition). Additionally, the procedure has the option remember indicated, forcing Maple to keep track of which values of the function have already been found, so that these can be directly looked up, instead of having to be re-computed.

```
>  seq(f2(i), i=1..15);
```
$$1, 1, 2, 3, 5, 8, 13, 21, 34, 55, 89, 144, 233, 377, 610$$

Now, to illustrate the difference in computational complexity, we shall compare the procedural and -> methods using the time function of Maple:

```
>  st:=time():seq(f1(i), i=1..20):time()-st;
```
$$2.050$$

```
>  st:=time():seq(f2(i), i=1..100):time()-st;
```
$$.017$$

So, it is clear that the `remember` option can make an enormous difference in time complexity.

Another way to improve the efficiency of a recursively defined function is to rewrite it to avoid the use of recursion. Instead, we rework it so that it uses an iterative algorithm. In constructing the iterative algorithm, the key components are to create a form of loop (either a `for` or a `while` loop in Maple) that will compute values starting from the smallest values, upward. This method of programming is called `bottom up`: where the smallest values of a sequence are computed and then used for the larger values.

```
> IterFib:=proc(n::integer)
>     local x,y,z,i;
>     if n=1 then y:=1;
>     else x:=1; y:=1;
>         for i from 2 to n-1 do
>             z:=x+y;
>             x:=y; y:=z;
>         od;
>     fi;
> y;
> end:
```

Contrast this with the recursive procedure `f2` which we defined earlier.

```
> eval(f2);
```
$$\mathbf{proc}(n::integer)$$
$$\mathbf{option}\ remember;$$
$$\mathbf{if}\ n \leq 2\ \mathbf{then}\ \text{RETURN}(1)\ \mathbf{fi};\ \text{f2}(n-1) + \text{f2}(n-2)$$
$$\mathbf{end}$$

Both the base cases and the recursive step are explicitly stated in the procedure body. The algorithm first attempts to compute the actual value directly, and asks for the values of sub-cases as required. This method of programming is know as "top down" for this reason: larger values are computed by breaking the input into smaller parts and combining results, similar to traversing down a binary tree.

Note that the recursive procedure with option remember and the iterative procedure perform about the same. For the first twenty Fibonacci numbers, we obtain

```
> st:=time():seq(RecFib(j), j=1..100):time()-st;
```
$$0$$

which is quite comparable to the times we obtained for `f2`.

Note that the purely recursive implementation **f1** cannot possibly be used compute **f2(100)**. In fact, it a good exercise is to show that to do so, **f2** would need to be invoked approximately

> f2(99);

$$218922995834555169026$$

times in order to handle all of the subcases which arise. Even at a billion subcases per second, this would require more than 6000 years to complete.

3.4   Computations and Explorations

In this section of material, we will explore how Maple can be used to solve Questions 4, 5 and 8 of the "Computations and Explorations" section of the textbook.

4. How many twin primes can you find?

   Solution: To determine how many twin primes there are, we will use the numtheory package of Maple, which contains the functions `nextprime`, `prevprime` and `ithprime`

   ```
   > with(numtheory):
   > list_of_primes:=[seq(ithprime(i), i=1..50)];
   ```
   $list\_of\_primes := [2, 3, 5, 7, 11, 13, 17, 19, 23, 29, 31, 37, 41, 43, 47, 53,$
   $59, 61, 67, 71, 73, 79, 83, 89, 97, 101, 103, 107, 109, 113, 127, 131, 137,$
   $139, 149, 151, 157, 163, 167, 173, 179, 181, 191, 193, 197, 199, 211,$
   $223, 227, 229]$

   Now, having formed a list of primes, we wish to extract any twin primes that occur in this list of primes.

   ```
   > twinprime_list:=[]:
   > for i from 1 to nops(list_of_primes)-1 do
   > if list_of_primes[i+1]-list_of_primes[i]=2 then
   > twinprime_list
   >    :=[op(twinprime_list),
   >      [list_of_primes[i], list_of_primes[i+1]
   >        ]];
   > fi;
   > od;
   > twinprime_list;
   ```
   $[[3, 5], [5, 7], [11, 13], [17, 19], [29, 31], [41, 43], [59, 61], [71, 73],$
   $[101, 103], [107, 109], [137, 139], [149, 151], [179, 181], [191, 193],$
   $[197, 199], [227, 229]]$

Now, instead of outputting the prime pairs, the sequence number of primes may indicate a pattern. So, we will construct a list of the 'i's that are "twinned" together.

```
>  ilist:=[]:
>  for i from 1 to nops(list_of_primes)-1 do
>      if list_of_primes[i+1]-list_of_primes[i] = 2 then
>          ilist:=[op(ilist), [i, i+1]];
>      fi;
>  od;
>  ilist;
```

$$[[2,3],[3,4],[5,6],[7,8],[10,11],[13,14],[17,18],[20,21],[26,27],$$
$$[28,29],[33,34],[35,36],[41,42],[43,44],[45,46],[49,50]]$$

It appears that there is no obvious pattern occurring.

5. Determine which Fibonacci numbers are divisible by 5, which are divisible by 7, and which are divisible by 11. Prove that your conjectures are correct.

   Solution: First, we shall generate some data to work with.

```
>  with(combinat): # get correct definition of 'fibonacci'
>  fib_list := [seq([n, fibonacci(n)], n = 0..50)]:
```

We want to determine those indices n for which the nth Fibonacci number is divisible by 5. One way to do this is to construct a list, by testing the data above, and adding to the list only those indices n for which the test returns true.

```
>  mult5 := NULL;
```

$$mult5 :=$$

```
>  for u in fib_list do
>      if op(2, u) mod 5 = 0 then
>          mult5 := mult5, op(1, u);
>      fi;
>  od;
>  mult5 := [mult5];
```

$$mult5 := [0, 5, 10, 15, 20, 25, 30, 35, 40, 45, 50]$$

This constructs a list indicating which among the first 50 Fibonacci numbers are multiples of 5. The data indicate the the nth Fibonacci number F_n is divisible by 5 only if n is.

To obtain evidence for the converse, we should test whether F_{5n} is divisible by 5, for as many n as possible. To make our test concise, and yet allow for testing a large range of values, we'll design it so that no output is produced unless a counterexample is found.

```
>    for n from 1 to 100 do
>      if fibonacci(5 * n) mod 5 <> 0 then
>        printf('The %dth Fibonacci number %d ',
>               5 * n, fibonacci(5 * n));
>        printf('(n = %d) is a counterexample\n', n);
>      fi;
>    od;
```

Hence, there is no counterexample among the first 500 Fibonacci numbers. You can try this with values larger than 100 also, to gain further evidence.

Another, slightly different approach can be used to locate the Fibonacci numbers divisible by a given integer, here 7. We simply build the divisibility test into the command to generate the data.

```
>    fib_list := seq([n, fibonacci(n) mod 7], n = 1..50);
```
$$fib\_list := [1, 1], [2, 1], [3, 2], [4, 3], [5, 5], [6, 1], [7, 6], [8, 0], [9, 6],$$
$$[10, 6], [11, 5], [12, 4], [13, 2], [14, 6], [15, 1], [16, 0], [17, 1], [18, 1],$$
$$[19, 2], [20, 3], [21, 5], [22, 1], [23, 6], [24, 0], [25, 6], [26, 6], [27, 5],$$
$$[28, 4], [29, 2], [30, 6], [31, 1], [32, 0], [33, 1], [34, 1], [35, 2], [36, 3],$$
$$[37, 5], [38, 1], [39, 6], [40, 0], [41, 6], [42, 6], [43, 5], [44, 4], [45, 2],$$
$$[46, 6], [47, 1], [48, 0], [49, 1], [50, 1]$$

We can now select the indices of those pairs whose second member is equal to 0.

```
>    mult7 := NULL:
>    for u in fib_list do
>      if op(2, u) = 0 then
>        mult7 := mult7, op(1, u);
>      fi;
>    od;
>    mult7 := [mult7];
```
$$mult7 := [8, 16, 24, 32, 40, 48]$$

We can begin to notice a pattern in this data as follows.

```
>    map(x -> x / 8, mult7);
```
$$[1, 2, 3, 4, 5, 6]$$

You can try to verify that this pattern persists by replacing 50 in the definition of `fib_list` by much larger numbers.

The tests for divisibility by 11 we leave to you.

8. The notorious $3x + 1$ conjecture (also known as *Collatz' Conjecture* and by many other names) states that no matter which integer x you start with, iterating the function $f(x)$, where $f(x) = x/2$ if x is even and

$f(x) = 3x + 1$ if x is odd, always produces the integer 1. Verify this conjecture for as many positive integers as possible.

Solution: To begin, we need to define the function we shall be examining.

```
>   Collatz := proc(n::integer)
>     if type(n, even) then
>       n / 2;
>     else
>       3 * n + 1;
>     fi;
>   end:
```

Now we write a function that will iterate the Collatz function until the value obtained is equal to 1. We include a count variable for two reasons: First, we want to get some idea of how long it takes for the iterates to stabilize; second, since we don't know for certain that the iterates will stabilize for a given value of the input seed, we code an upper limit (large) on the number of iterates to compute.

```
>   IC := proc(seed::integer)
>     local sentinel, count;
>
>     count := 0;
>     sentinel := seed;
>     while sentinel <> 1 and count < 1000^1000 do
>       sentinel := Collatz(sentinel);
>       count := count + 1;
>     od;
>     RETURN(count);
>   end:
```

To verify the conjecture for the first 100 integers, we can use our function IC as follows.

```
>   seq(IC(i), i = 1..100);
```
$0, 1, 7, 2, 5, 8, 16, 3, 19, 6, 14, 9, 9, 17, 17, 4, 12, 20, 20, 7, 7, 15, 15, 10,$
$23, 10, 111, 18, 18, 18, 106, 5, 26, 13, 13, 21, 21, 21, 34, 8, 109, 8, 29,$
$16, 16, 16, 104, 11, 24, 24, 24, 11, 11, 112, 112, 19, 32, 19, 32, 19, 19,$
$107, 107, 6, 27, 27, 27, 14, 14, 14, 102, 22, 115, 22, 14, 22, 22, 35, 35, 9,$
$22, 110, 110, 9, 9, 30, 30, 17, 30, 17, 92, 17, 17, 105, 105, 12, 118, 25,$
$25, 25$

Note that the fact that the function eventually *stopped* is the verification that we sought.

3.5   Exercises/Projects

Exercise 1. Use Maple to find and prove formulae for the sum of the first k nth powers of positive integers for $n = 4, 5, 6, 7, 8, 9$, and 10.

Exercise 2. Use Maple to study the McCarthy 91 function. (See Page 227 of the text).

Exercise 3. Write a Maple procedure to find the smallest (that is, the *first*) consecutive sequence of n composite positive integers, for an arbitrary positive integer n.

Exercise 4. Use Maple to develop a procedure for generating Ulam numbers (defined on Page 226 of the text). Make and numerically study conjectures about the distribution of these numbers.

Exercise 5. Write a Maple procedure that takes an integer k as input, and determines whether or not the product of the first k primes, plus 1, is prime or not, by factoring this number.

Exercise 6. Another way to show that there are infinitely many primes is to assume that there are only n primes $p_1, p_2, \ldots, p_n$. but this is a contradiction since $p_1 p_2 \cdots p_n + 1$ has at least one prime factor and it is not divisible by p_i, $i = 1, 2, \ldots, n$. Find the smallest prime factor of $2 \cdot 3 \cdots p_n + 1$ for all positive integers n not exceeding 20. For which n is this number prime.

Exercise 7. The *Lucas numbers* satisfy the recurrence $L_n = L_{n-1} + L_{n-2}$ and the initial conditions $L_0 = 2$ and $L_1 = 1$. Use Maple to gain evidence for conjectures about the divisibility of Lucas numbers by different integer divisors.

Exercise 8. A sequence $a_1, a_2, a_3, \ldots$ is called *periodic* if there are positive integers N and p for which $a_n = a_{n+p}$, for all $n \geq N$. The least integer p for which this is true is called the *period* of the sequence. The sequence $a_1, a_2, a_3, \ldots$ is said to be *periodic modulo* m, for a positive integer m, if the sequence $a_1 \pmod{m}, a_2 \pmod{m}, a_3 \pmod{m}, \ldots$ is periodic. Use Maple to determine whether the Fibonacci sequence is periodic modulo m, for various integers m and, if so, find the period. Can you, by examining enough different values of m, make any conjectures concerning the relationship between m and the period? Do the same thing for other sequences that you find interesting.

4 Counting

Counting is fundamental to the study of discrete mathematics, the complexity of algorithms, combinatorics, and some branches of algebra such as finite group theory. This chapter presents a variety of techniques that are available in Maple for counting a diverse collection of discrete objects, including combinations and permutations of finite sets. Objects can be counted by using formulae or other algorithms, or by listing them and noting directly the size of the list. The latter approach is facilitated by a number of Maple procedures that can be used to generate combinatorial structures.

Most of the Maple procedures relevant to this chapter dwell in one of two packages. The `combinat` package is a standard part of the Release 3 Maple library.

A new package `combstruct` is available as a share library for MapleV, Release 3, and is a standard package in Release 4. You can access the services provided by either of these packages by using the `with` command to load it into your Maple session. (If you are using Maple V, Release 3, you must also enter `with(share)` before typing `with(combstruct)`).

It is useful to know that the `combstruct` package, while providing a great variety of procedures, has organized some of the basic functions into groups related to a particular combinatorial object (such as, for example, combinations, or partitions). For many types of combinatorial objects, there are Maple procedures to do the following operations.

1. You can construct all objects of that type associated with a given integer. The procedure to do this is usually given a name reflecting the type of object. (For example, `permute` and `partitions`.)
2. You can count all objects of that type associated with a given integer. These procedures generally begin with the string "numb" and are completed by an abbreviation of the type of object being counted. (For example, `numbperm` and `numbpart`.)
3. You can generate a random object of that type associated with a given integer. An abbreviation of the type of object being generated, prefixed with the string "rand" is how these routines are normally named. (For example, `randperm` and `randpart`.)

Of course, there are many other functions that do not fit this scheme, as well.

4.1   Relevant Maple functions

The `combinat` package contains many functions pertaining to counting
and generating combinatorial structures. The list of functions in this package is:

```
> with(combinat);
```
  [Chi, *bell*, binomial, *cartprod, character, choose, composition, conjpart,*
    decodepart, encodepart, fibonacci, firstpart, graycode, inttovec, lastpart,
    multinomial, nextpart, numbcomb, numbcomp, numbpart, numbperm,
    partition, permute, powerset, prevpart, randcomb, randpart, randperm,
    stirling1, stirling2, subsets, vectoint]

There is another package, `combstruct`, available in Maple V, Release 4,
which also deals with combinatorial structures. Most of what this package
does is beyond the scope of this book, but some of its functions expand
what the `combinat` package does. The `combstruct` package supplies functions

`count:`	- to count the number of objects of a given size
`draw:`	- to generate a random object of a given size
`allstructs:`	- to generate all the objects of a given size
`iterstructs:`	- to generate the "next" structure of a given size. The rele-vant structures that `combstruct` can deal with are Permu-tation, Combination/Subset, Partition. Since not everyone will have access to Release 4 of Maple, more time will be spent using the `combinat` package.

To access the services provided by the `combstruct` package, type

```
> with(combstruct);
```
          [*allstructs, count, draw, finished, iterstructs, nextstruct*]

If you are using Release 3 of Maple, you will *first* need to issue the command `with(share)`, since the `combstruct` package is a part of the share
library in Release 3.

The functions in the `combinat` package for combinations are `numbcomb`,
`choose`, and `randcomb`. This is the number of ways of choosing 2 fruits
from an apple, an orange and a pear.

```
> numbcomb([apple, orange, pear], 2);
```
                                     3

Here are the possible choices:

```
> choose([apple, orange, pear], 2);
```
$$[[apple, orange], [apple, pear], [orange, pear]]$$

The function numbcomb counts the number of combinations (or r-combinations) of a set. The function choose lists the actual combinations. Thus there will always be numbcomb elements listed by choose.

```
> nops(");
```
$$3$$

And if we have 2 apples and no pear (an example with indistinguishable elements),

```
> numbcomb([apple, apple, orange],2);
```
$$2$$

With the choices:

```
> choose([apple, apple, orange],2);
```
$$[[apple, apple], [apple, orange]]$$

If we don't supply the second argument, all possible combinations of all possible sizes are considered.

```
> numbcomb([apple, apple, orange]);
```
$$6$$

```
> choose([apple, apple, orange]);
```
$$[[], [apple], [orange], [apple, orange], [apple, apple],$$
$$[apple, apple, orange]]$$

We can also choose combinations at random.

```
> randcomb([chocolate, vanilla, cookiedough],2);
```
$$[vanilla, cookiedough]$$

```
> randcomb(5,3);
```
$$\{2, 4, 5\}$$

In this example, the 5 represents the set $\{1, 2, 3, 4, 5\}$.

Using combstruct, we would solve the above problems as follows:

```
> count(Combination([apple,orange,pear]),size=2);
```
$$3$$

```
>  allstructs(Combination([apple,orange,pear]), size=2);
                   [[apple, orange], [apple, pear], [orange, pear]]

>  draw(Combination([chocolate,vanilla,cookiedough]),size=2);
                          [chocolate, vanilla]
```

Binomial coefficients can be calculated either by calling the `numbcomb` function with an integer as the first argument,

```
>  numbcomb(10,5);
```
$$252$$

or, we can calculate $C(n, r)$, using the `binomial` function. So we solve Example 7 in section 4.3 as follows:

```
>  binomial(10,5);
```
$$252$$

When n and r are non-negative integers and $r \leq n$, `binomial` and `numbcomb` behave identically. The procedure `binomial` is more general, and extends the definition of the binomial coefficients. We shall not discuss its more general use here.

4.2   More Combinatorial Functions

In this section we discuss some combinatorial functions, useful in counting, that arise as coefficients of certain polynomials.

Binomial Coefficients

The binomial coefficients are the coefficients of the polynomial $(a + b)^n$ when it is expanded.

```
>  for n from 1 to 7 do
>    sort(expand((a + b)^n));
>  od;
```

$$a + b$$
$$a^2 + 2\,a\,b + b^2$$
$$a^3 + 3\,a^2\,b + 3\,a\,b^2 + b^3$$
$$a^4 + 4\,a^3\,b + 6\,a^2\,b^2 + 4\,a\,b^3 + b^4$$
$$a^5 + 5\,a^4\,b + 10\,a^3\,b^2 + 10\,a^2\,b^3 + 5\,a\,b^4 + b^5$$
$$a^6 + 6\,a^5\,b + 15\,a^4\,b^2 + 20\,a^3\,b^3 + 15\,a^2\,b^4 + 6\,a\,b^5 + b^6$$
$$a^7 + 7\,a^6\,b + 21\,a^5\,b^2 + 35\,a^4\,b^3 + 35\,a^3\,b^4 + 21\,a^2\,b^5 + 7\,a\,b^6 + b^7$$

These numbers can be accessed directly in Maple using the `binomial` function from the Maple library.

```
>    for n from 1 to 7 do
>       seq(binomial(n, k), k = 0..n);
>    od;
```

$$1, 1$$
$$1, 2, 1$$
$$1, 3, 3, 1$$
$$1, 4, 6, 4, 1$$
$$1, 5, 10, 10, 5, 1$$
$$1, 6, 15, 20, 15, 6, 1$$
$$1, 7, 21, 35, 35, 21, 7, 1$$

The value of `binomial(n, k)` is the coefficient of the binomial term $a^k b^{n-k}$ (which is equal to the coefficient of $a^{n-k} b^k$) in the expansion of $(a + b)^n$.

Given numeric arguments, `binomial` evaluates to a number.

```
>    binomial(100,53);
```

$$84413487283064039501507937600$$

However, if a symbolic argument is given, `binomial` returns unevaluated.

```
>    n := 'n': # clear values
>    k := 'k': # from n and k
>    binomial(n, 9);
```

$$\mathrm{binomial}(n, 9)$$

You can express this as a rational function of the variable **n** by calling `expand`.

```
>    expand(");
```

$$\frac{1}{362880} n^9 - \frac{1}{10080} n^8 + \frac{13}{8640} n^7 - \frac{1}{80} n^6 + \frac{1069}{17280} n^5 - \frac{89}{480} n^4 + \frac{29531}{90720} n^3$$
$$- \frac{761}{2520} n^2 + \frac{1}{9} n$$

However, this only works if at most one of the arguments is symbolic.

```
>    binomial(n, k);
```

$$\mathrm{binomial}(n, k)$$

```
>    expand(");
```

$$\mathrm{binomial}(n, k)$$

To determine the definition, in terms of factorials, you can use the multi-faceted `convert` command.

```
>   convert(binomial(n, k), factorial);
```
$$\frac{n!}{k!\,(n-k)!}$$

The `convert` procedure is a general purpose conversion utility that can be used to transform expressions from one form to another, equivalent form. Here, it transforms a symbolic expression involving the call to the procedure `binomial` to an equivalent on expressed using factorials. Because `convert` accepts a wide variety of argument types, its documentation is spread over many of the online help pages. But a good place to start to find out more about `convert`, is the main help page for this command, accessed by typing `?convert`. This facility can be used to prove combinatorial identities involving the binomial coefficients. A little care is needed, however, to take into account the degree of evaluation that is performed at each step, lest things that are equal not be recognized as such. For example, the famous identity $C(n,k) = C(n,n-k)$ can be proved as follows.

```
>   left := binomial(n, k);
```
$$left := \text{binomial}(n,k)$$

```
>   right := binomial(n, n - k);
```
$$right := \text{binomial}(n,n-k)$$

We want to prove that `left` and `right` are equal. Note that

```
>   evalb(left = right);
```
$$false$$

This occurs because `left` and `right` have been evaluated insufficiently so far. To overcome this lack of recognition, we use `convert`.

```
>   left := convert(left, factorial);
```
$$left := \frac{n!}{k!\,(n-k)!}$$

```
>   right := convert(right, factorial);
```
$$right := \frac{n!}{k!\,(n-k)!}$$

```
>   evalb(left = right);
```
$$true$$

There is often a certain amount of guesswork involved in coercing symbolic expressions into a form that is useful for a given problem. Maple is
designed to allow you to easily experiment with expressions, so that you
can discover the "right" form for a particular application.

Multinomial Coefficients

For computing the number of permutations of a finite set in which some
members are indistinguishable from others (such a "set" is usually called a
"multiset"), Maple provides the procedure `multinomial` in the `combinat`
package. It calculates the multinomial coefficients, that is, numbers of the
form

$$\frac{n!}{n_1! n_2! n_3! \cdots n_k!}$$

in which $n_1, n_2, n_3, \ldots, n_k$ are non-negative integers whose sum is n. The
first argument to `multinomial` is the integer n, while the remaining arguments are the numbers $n_1, n_2, \ldots, n_k$ from the denominator.

For example, let us compute the number of distinct strings obtained by
permuting the letters of the word "MISSISSIPPI" (a classical example).
Here, there is 1 "M", and there are 4 "I"s, 4 "S"s, and 2 "P"s. This gives
a total of 11 characters. Hence, the number of distinct strings is

```
>   combinat[multinomial](11, 1, 4, 4, 2);
                  34650
```

Note that the first argument must be the sum of the remaining arguments;
if not, an error is indicated.

```
>   combinat[multinomial](11, 1, 4, 4, 3);

Error, (in combinat[multinomial])
1st argument must equal the sum of the others
```

The multinomial coefficient displayed above is called a "coefficient" be-
cause it is the coefficient of the multinomial $x_1^{n_1} x_2^{n_2} \cdots x_k^{n_k}$ in the expansion
of the polynomial $(x_1 + x_2 + \cdots + x_k)^n$. We can see some examples of this
using Maple. (We'll use variables a, b, c, and so on, since they are easier
to read than x1, x2, x3, etc.)

```
>   p := (a + b + c)^5;
```
$$p := (a + b + c)^5$$

```
>   p := expand(p);
```
$$p := 5 a^4 b + 10 a^3 b^2 + 10 a^2 b^3 + 5 a b^4 + a^5 + b^5 + c^5 + 5 a c^4 + 5 a^4 c$$
$$+ 10 a^3 c^2 + 10 a^2 c^3 + 5 b c^4 + 5 b^4 c + 10 b^3 c^2 + 10 b^2 c^3 + 20 a^3 b c$$
$$+ 30 a^2 b^2 c + 30 a^2 b c^2 + 20 a b^3 c + 30 a b^2 c^2 + 20 a b c^3$$

There is a function `coeff` that extracts the coefficient of a variable in a polynomial.

```
> coeff(x^3 - 5*x^2 + 2, x^2);
```
$$-5$$

```
> coeff(x^3 - 5*x^2 + 2, x);
```
$$0$$

However, it only works with univariate polynomials. You can, however, access the individual multinomials in a multivariate polynomial, using the op command.

```
> op(3, p);
```
$$10\,a^2\,b^3$$

```
> op(p);
```
$$5\,a^4\,b, 10\,a^3\,b^2, 10\,a^2\,b^3, 5\,a\,b^4, a^5, b^5, c^5, 5\,a\,c^4, 5\,a^4\,c, 10\,a^3\,c^2, 10\,a^2\,c^3,$$
$$5\,b\,c^4, 5\,b^4\,c, 10\,b^3\,c^2, 10\,b^2\,c^3, 20\,a^3\,b\,c, 30\,a^2\,b^2\,c, 30\,a^2\,b\,c^2, 20\,a\,b^3\,c,$$
$$30\,a\,b^2\,c^2, 20\,a\,b\,c^3$$

This, unfortunately, depends upon the ordering of the multinomials in the polynomial p making it impossible to predict *which* among the multinomials in p is going to be extracted. To get around this problem, use the sort command first.

```
> p := sort(p);
```
$$p := a^5 + 5\,a^4\,b + 5\,a^4\,c + 10\,a^3\,b^2 + 20\,a^3\,b\,c + 10\,a^3\,c^2 + 10\,a^2\,b^3$$
$$+ 30\,a^2\,b^2\,c + 30\,a^2\,b\,c^2 + 10\,a^2\,c^3 + 5\,a\,b^4 + 20\,a\,b^3\,c + 30\,a\,b^2\,c^2$$
$$+ 20\,a\,b\,c^3 + 5\,a\,c^4 + b^5 + 5\,b^4\,c + 10\,b^3\,c^2 + 10\,b^2\,c^3 + 5\,b\,c^4 + c^5$$

```
> op(3, p);
```
$$5\,a^4\,c$$

```
> terms := [op(p)];
```
$$terms := [a^5, 5\,a^4\,b, 5\,a^4\,c, 10\,a^3\,b^2, 20\,a^3\,b\,c, 10\,a^3\,c^2, 10\,a^2\,b^3,$$
$$30\,a^2\,b^2\,c, 30\,a^2\,b\,c^2, 10\,a^2\,c^3, 5\,a\,b^4, 20\,a\,b^3\,c, 30\,a\,b^2\,c^2, 20\,a\,b\,c^3,$$
$$5\,a\,c^4, b^5, 5\,b^4\,c, 10\,b^3\,c^2, 10\,b^2\,c^3, 5\,b\,c^4, c^5]$$

The multinomials are sorted lexicographically. (For example, $a^3 < a^2b < \cdots < b^3$.) To repair the deficiency in `coeff` that prevents it from handling multivariate polynomials, we can write our own routine, `mcoeff` that does this job for us. Since `coeff` is implemented in the Maple kernel, it is not possible for a user to redefine its behavior, so that a separate routine is necessary. For simplicity, our `mcoeff` procedure will only handle polynomials with numerical coefficients. The algorithm used here is as follows.

1. input a polynomial p and a multinomial term `term`.

2. process p as follows:

  (a) sort p into q

  (b) create a list r of the multinomial terms in q.

  (c) create a multiset m consisting of the multinomials in q with multiplicity equal to its coefficient. (Note that this is not a "true" multiset, as the coefficient may be negative or nonintegral.)

3. search the list m for an entry matching `term` and, if found, return the coefficient. Otherwise, return 0.

Here, then, is the Maple code for `mcoeff`.

```
>   mcoeff := proc(p::polynom, term::polynom)
>      local m, # list of multinomials
>            t, # index into m
>            x, # dummy variable
>            q, # sorted input
>            r; # multiset of multinomials and coefficients
>      q := sort(p); r := [op(q)];
>      m := map(x -> [coeffs(x), x / coeffs(x)], r);
>      for t in m do
>        if term = op(2, t) then RETURN(op(1, t)); fi;
>      od;
>      RETURN(0);
>   end:
```

For example, to locate the coefficient of a^2b^3 in the multivariate polynomial $(a + b + x)^5$, we can use `mcoeff` as follows.

```
>   p := (a + b + c)^5;
```

$$p := (a + b + c)^5$$

```
>   p := expand(p);
```

$$p := 5\,a\,c^4 + 10\,a^3\,b^2 + c^5 + 10\,b^2\,c^3 + 5\,b^4\,c + 10\,b^3\,c^2 + 10\,a^3\,c^2$$
$$+ 10\,a^2\,b^3 + 5\,a^4\,b + 5\,a^4\,c + 10\,a^2\,c^3 + 5\,a\,b^4 + 5\,b\,c^4 + 20\,a\,b\,c^3$$
$$+ 20\,a\,b^3\,c + 30\,a\,b^2\,c^2 + 20\,a^3\,b\,c + 30\,a^2\,b^2\,c + 30\,a^2\,b\,c^2 + a^5 + b^5$$

```
>   mcoeff(p, a^2 * b^3);
```

Asking for the coefficient of a multinomial not in the polynomial results in zero.

```
> mcoeff(p, x^5);
```
$$0$$

If the input polynomial p is a polynomial in a single variable, then the call mcoeff(p, x^n) is equivalent to the call coeff(p, x^n) or coeff(p, x, n). (A calling syntax in the latter style is not supported by mcoeff.)

```
> mcoeff(x^3 - 2*x^2 + 1, x^2);
```
$$-2$$

```
> coeff(x^3 - 2*x^2 + 1, x^2);
```
$$-2$$

```
> coeff(x^3 - 2*x^2 + 1, x, 2);
```
$$-2$$

The routine mcoeff provides another means by which we can determine multinomial coefficients. For example

```
> with(combinat):
> multinomial(6, 1, 2, 3);
```
$$60$$

```
> p := expand((a + b + c)^6);
```
$$p := 6\,a^5\,c + 6\,a^5\,b + 15\,a^4\,c^2 + 15\,a^4\,b^2 + 20\,a^3\,c^3 + 20\,a^3\,b^3 + 15\,a^2\,c^4$$
$$+ 15\,a^2\,b^4 + 6\,a\,c^5 + 6\,a\,b^5 + 6\,b\,c^5 + 15\,c^4\,b^2 + 20\,c^3\,b^3 + 15\,c^2\,b^4$$
$$+ 6\,c\,b^5 + a^6 + b^6 + 30\,a^4\,c\,b + 60\,a^3\,c^2\,b + 60\,a^3\,c\,b^2 + 60\,a^2\,c^3\,b$$
$$+ 90\,a^2\,c^2\,b^2 + 60\,a^2\,c\,b^3 + 30\,a\,c^4\,b + 60\,a\,c^3\,b^2 + 60\,a\,c^2\,b^3 + 30\,a\,c\,b^4$$
$$+ c^6$$

```
> mcoeff(p, a * b^2 * c^3);
```
$$60$$

Stirling Numbers

Another combinatorially significant set of numbers that arises as the set of coefficients of special polynomials is the set of Stirling numbers. The Sterling polynomial of degree n is defined to be

$$S_n(x) = x \cdot (x - 1) \cdot (x - 2) \cdots (x - n + 1).$$

When expanded, $S_n(x)$ takes the form

$$S_n(x) = s(n, 1)x + s(n, 2)x^2 + s(n, 3)x^3 + \cdots + s(n, n)x^n.$$

The coefficients $s(n, k)$, for $1 \leq k \leq n$, are called the Stirling numbers (of the first kind).

We can use Maple to generate the Stirling polynomials as follows.

```
>  n := 'n'; i := 'i';
```
$$n := n$$
$$i := i$$

```
>  S(n) := product(x - i, i = 0..n-1);
```
$$S(n) := \frac{(-1)^n \, \Gamma(n - x)}{\Gamma(-x)}$$

This expression Maple insists on displaying with the use of the Gamma function Γ. The Gamma function is a continuous extension of the factorial function to real numbers. For a non-negative integer n, we have $\Gamma(n+1) = n!$. But, for specific values of n, we can coerce Maple into representing the Stirling polynomials *as polynomials*, by using `simplify`.

```
>  subs(n = 9, S(n));
```
$$-\frac{\Gamma(9 - x)}{\Gamma(-x)}$$

```
>  simplify(");
```
$$x \, (x - 1) \, (x - 2) \, (x - 3) \, (x - 4) \, (x - 5) \, (x - 6) \, (x - 7) \, (x - 8)$$

```
>  expand(");
```
$$x^9 - 36 \, x^8 + 546 \, x^7 - 4536 \, x^6 + 22449 \, x^5 - 67284 \, x^4 + 118124 \, x^3$$
$$- 109584 \, x^2 + 40320 \, x$$

```
>  sort(");
```
$$x^9 - 36 \, x^8 + 546 \, x^7 - 4536 \, x^6 + 22449 \, x^5 - 67284 \, x^4 + 118124 \, x^3$$
$$- 109584 \, x^2 + 40320 \, x$$

```
>  coeffs(");
```
$$22449, 118124, -109584, -36, -67284, -4536, 546, 1, 40320$$

```
>  ["];
```
$$[22449, 118124, -109584, -36, -67284, -4536, 546, 1, 40320]$$

Thus, we have a list of the Stirling numbers $s(9, k)$, for $k = 1, 2, \ldots, 9$.

You can access the Stirling numbers directly in Maple, using the function `stirling1` in the `combinat` package.

```
>   with(combinat):
>   for n from 1 to 7 do
>     seq(stirling1(n,i), i = 1..n);
>   od;
```

$$1$$
$$-1, 1$$
$$2, -3, 1$$
$$-6, 11, -6, 1$$
$$24, -50, 35, -10, 1$$
$$-120, 274, -225, 85, -15, 1$$
$$720, -1764, 1624, -735, 175, -21, 1$$

There are some interesting patterns in the resulting triangle. Try to compute more Stirling numbers and see if you can make any conjectures about the patterns that you see.

4.3 Permutations

We have already shown how to count and generate combinations using Maple. We shall now introduce analogous capabilities of Maple for working with permutations. The corresponding Maple functions for permutations are `numbperm`, `permute` and `randperm`. Since all reside in the `combinat` package, it must be loaded before they can be used.

```
>   with(combinat):
>   numbperm([S,U,C,C,E,S,S]);
```
$$420$$

```
>   permute([a,b,c]);
```
$$[[a, b, c], [a, c, b], [b, a, c], [b, c, a], [c, a, b], [c, b, a]]$$

```
>   randperm([S,U,C,C,E,S,S]);
```
$$[S, C, E, S, S, C, U]$$

```
>   randperm(5);
```
$$[4, 3, 1, 5, 2]$$

Using the `combstruct` package, these examples are done as follows:

```
>   with(combstruct):
>   count(Permutation([S,U,C,C,E,S,S]));
```
$$420$$

```
>  allstructs(Permutation([a,b,c]));
```
$$[[a, b, c], [a, c, b], [b, a, c], [b, c, a], [c, a, b], [c, b, a]]$$

```
>  draw(Permutation(5));
```
$$[3, 2, 5, 1, 4]$$

The function **subsets** allows you to generate all the subsets of a given set. Since subsets and combinations are just different names for the same thing, you can use this function to generate combinations. The function **subsets** returns a table that contains two entries. One is called **nextvalue**, and is a procedure to generate the next combination, and the other is **finished**, a true/false flag that tells you when they have all been generated.

```
>  S := combinat[subsets]({a,b}):
>  while not S[finished] do
>    S[nextvalue]();
>  od;
```
$$\{\}$$
$$\{a\}$$
$$\{b\}$$
$$\{a, b\}$$

Using **combstruct**, one does the same thing using the **iterstructs** function. The procedure **iterstructs** also returns a table, but this time one uses the functions **next** and **finished** to iterate.

```
>  S := iterstructs(Subset({a,b})):
>  while not finished(S) do
>    nextstruct(S);
>  od;
```
$$\{\}$$
$$\{a\}$$
$$\{b\}$$
$$\{a, b\}$$

Using **iterstructs**, we can also iterate over permutations and partitions. In addition, we can specify which size of object we want to see.

```
>  P := iterstructs(Permutation([a,b,b]), size=2):
>  while not finished(P) do
>    nextstruct(P);
>  od;
```
$$[a, b]$$
$$[b, a]$$
$$[b, b]$$

Because Maple's permutation functions can solve problems of permutation with indistinguishable elements as easily as without indistinguishable elements, some of the text exercises become trivial. For example, Exercise 26 says "How many different strings can be made from the letters in MISSISSIPPI using all the letters". The solution can be found in one step:

```
> numbperm([M,I,S,S,I,S,S,I,P,P,I]);
                        34650
```

Question 29 is similar, but involves some extra steps. It asks "How many different strings can be made from the letters in ORONO, using some or all of the letters?" To find the solution, we first calculate the number of 1-permutations, then 2-permutations etc.

```
> total := 0:
> for i from 1 to 5 do
>        total := total + numbperm([O,R,O,N,O],i);
> od:

> total;
                          63
```

There are 63 possible strings using some or all of the letters in ORONO. 64 if we count the string with 0 letters.

```
> numbperm([O,R,O,N,O],0);
                           1
```

Using the `combstruct` package, we can find the answer in one step.

```
> with(combstruct):
> count(Permutation([O,R,O,N,O]), size='allsizes');
                          66
```

However, most of this section involves thinking and understanding the question. Maple can help calculate the numbers of combinations and permutations, but it is up to you to decide which values you need to calculate to reach the answer.

Partitions of Integers

There are also functions to do integer partitions. (An integer partition is a way of writing an integer n as a sum of positive integers, where order does not matter. So $5 = 1 + 1 + 3$ is one integer partition of 5.) Along with `numbpart`, `partition` and `randpart`, there are functions to generate partitions, one at a time, based on a given canonical order. All these functions are part of the `combinat` package which must, consequently, be loaded before you can access them.

```
>  with(combinat):
```

The number of partitions of a given integer can be counted using the procedure `numbpart`.

```
>  seq(numbpart(i), i = 1..20);
```
$$1, 2, 3, 5, 7, 11, 15, 22, 30, 42, 56, 77, 101, 135, 176, 231, 297, 385, 490,$$
$$627$$

The partitions of an integer can be computed using the `partition` function.

```
>  partition(5);
```
$$[[1, 1, 1, 1, 1], [1, 1, 1, 2], [1, 2, 2], [1, 1, 3], [2, 3], [1, 4], [5]]$$

This constructs the partitions of its argument as a list of lists, each sublist representing one partition.

As its name suggests, `randpart` simply creates a random partition of an integer.

```
>  randpart(20);
```
$$[1, 1, 1, 1, 1, 1, 2, 3, 3, 6]$$

Maple provides special functions for generating the sequence of all partitions of a given integer. Thus we have the routines `firstpart`, `nextpart`, `prevpart` and `lastpart`.

```
>  firstpart(4);
```
$$[1, 1, 1, 1]$$

```
>  nextpart(");
```
$$[1, 1, 2]$$

```
>  nextpart(");
```
$$[2, 2]$$

```
>  prevpart(");
```
$$[1, 1, 2]$$

```
>  nextpart("");
```
$$[1, 3]$$

```
>  lastpart(4);
```
$$[4]$$

4.4   Discrete Probability

To find the probability of an event in a finite sample space, one calculates
the number of times the event occurs, and divides by the total number of
possible outcomes (the size of the sample space).

As in Example 4, section 4.4, we calculate the probability of winning a
lottery, where we need to choose 6 numbers correctly out of 40 possible
numbers. The total number of ways to choose 6 numbers is

```
>   numbcomb(40,6);
```
$$3838380$$

and there is one winning combination. Thus the probability is

```
>   1/";
```
$$\frac{1}{3838380}$$

which we can see as a real number approximation by using the `evalf`
function — evaluation as a floating point number.

```
>   evalf(");
```
$$.2605265763 \; 10^{-6}$$

We could also force a decimal approximation of the result by using `1.0`,
or simply `1.`, to show that we wish to work with decimals instead of the
exact rational representation. For example, if we needed to choose from
50 numbers, the probability is

```
>   1./numbcomb(50,6);
```
$$.6292988981 \; 10^{-7}$$

For another example of the use of Maple in the study of discrete probability, let us use Maple to verify the assertion in Example 14 on Page 278 of
the text. The claim is that the expected value of the number of successes
for n Bernoulli trials, each with probability p of success, is np. We'll use
EV to denote expected value in Maple. (We cannot use E because that
symbol is reserved for the base of the natural logarithm.) We know that

```
>   p(X=k) := binomial(n, k) * p^k * (1 - p)^(n - k);
```
$$p(X = k) := binomial(n, k) \, p^k \, (1 - p)^{(n-k)}$$

From the definition we have

```
>   EV(X) := sum(k * p(X=k), k = 1..n);
```

$$EV(X) := \frac{(1 + \frac{p}{1-p})^n \, p \, n \, (1-p)^n}{(1-p)\,(1 + \frac{p}{1-p})}$$

```
>   simplify(");
```

$$n\,p$$

4.5   Generating Permutations and Combinations

Here is an implementation of the algorithm to generate the next r-combination (Example 5).

```
>   NextrCombination := proc(current, n, r)
>   local next, i, j;
>    # make a copy that we can change
>     next := table(current);
>     i := r;
>     while next[i] = n - r + i do i := i -1 od;
>     next[i] := next[i] + 1;
>     for j from i+1 to r do
>        next[j] := next[i] + j - i;
>     od;
>     [seq( next[i], i=1..r) ];  # return the answer
>   end:
```

Test it on the example.

```
>   NextrCombination([1,2,5,6], 6, 4);
```

$$[1, 3, 4, 5]$$

```
>   NextrCombination(",6,4);
```

$$[1, 3, 4, 6]$$

```
>   NextrCombination(",6,4);
```

$$[1, 3, 5, 6]$$

Some explanation is needed. First, the current combination is a list, not a set. This is because a list is ordered, but a set is unordered. To find the "next" combination, we need to know the order of elements in the current combination. But in Maple, the order we type a set and the order it appears inside Maple are not necessarily the same thing.

```
>   {pear, orange, apple};
```

$$\{pear, apple, orange\}$$

But it will always be the same for a list.

```
>   [pear,orange,apple];
```
$$[pear, orange, apple]$$

The next problem is that you cannot, before MapleV Release 4, assign a specific element inside a list.

```
>   mylist := [a,b,c,d]:
>   mylist[2] := e;
```
$$mylist_2 := e$$

So the first thing we do in this algorithm is make a table that contains all the elements in the combination. We can assign into a table, so our problem goes away.

```
>   mytable := table(mylist);
```
$$mytable := \text{table}([$$
$$3 = c$$
$$4 = d$$
$$1 = a$$
$$2 = e$$
$$])$$

```
>   mytable[2] := e;
```
$$mytable_2 := e$$

```
>   print(mytable);
```
$$\text{table}([$$
$$3 = c$$
$$4 = d$$
$$1 = a$$
$$2 = e$$
$$])$$

With the `combstruct` package, you can create an iterator which will produce all the objects of a certain size, one at a time.

```
>   it := iterstructs(Combination(6),size=4):
>   nextstruct(it);
```
$$\{3, 4, 1, 2\}$$

```
>   nextstruct(it);
```
$$\{3, 5, 1, 2\}$$

```
>   nextstruct(it);
```
$$\{3, 6, 1, 2\}$$

Calling this function a few more times leads us to:

```
>   nextstruct(it);
```
$$\{4, 5, 1, 2\}$$

where the next 4-combination is then:

```
>   nextstruct(it);
```
$$\{4, 6, 1, 2\}$$

by which we can see that this iterator is using the same lexicographic ordering as we used in Algorithm 3.

4.6   Computations and Explorations

3. (Computer Projects) Given a positive integer n, find the probability of selecting the six integers from the set $\{1, \dots, n\}$ that were mechanically selected in a lottery.

Solution: We will follow example 4 from the text. The total number of ways of choosing 6 numbers from n numbers is $C(n, 6)$, which is found with the procedure numbcomb in the combinat package. This gives us the total number of possibilities, only one of which will win.

```
>  Lottery := proc(n::posint)
>  local total;
>     total := combinat[numbcomb](n, 6);
>     1.0 / total;
>  end:
>
>  Lottery(49);
```
$$.7151123842 \; 10^{-7}$$

If the rules of the lottery change, so that the number of numbers chosen is something other than 6, then we must modify the procedure above. (For example, we may now choose 5 numbers from 49, instead of 6.) We can easily modify our program to allow us to specify how many numbers we want to choose, by adding another parameter.

```
>  Lottery2 := proc(n::posint, k::posint)
>  local total;
>     total := combinat[numbcomb](n,k);
>     1.0 / total;
>  end:
>  Lottery2(49,6);
```
$$.7151123842 \; 10^{-7}$$

```
>  Lottery(30,3);
```
$$.1684139615 \; 10^{-5}$$

11. Given positive integers n and r, list all the r-combinations, with repetition allowed, of the set $\{1, 2, \dots, n\}$.

Solution: The Maple function choose (in the combinat package), will list all the r-combinations of $\{1, 2, \dots, n\}$, but without repetitions. Thus we cannot use it directly. However, say we want all the 2-combinations of $\{1, 2, 3\}$, with repetitions. This means that along with $\{1, 2\}$, $\{2, 3\}$ and $\{1, 3\}$, we also want to include $\{1, 1\}$, $\{2, 2\}$ and $\{3, 3\}$. We want to be able choose each number up to 2 times. (We say that we can repeat element any number of times, but practically, since we can only choose 2 things in total, we only need to allow each number to appear at most 2 times.) So another way to look at the problem is to say we want all the 2-combinations, without repetition, of the set $\{1, 1, 2, 2, 3, 3\}$. In general, then, we can find all the r-combinations of $\{1, 2, \ldots, n\}$ with repetition by asking for all the r-combinations of $(1, \ldots, 1, 2, \ldots, 2, \ldots, n, \ldots, n)$, where each element appears r times.

```
>  RCombRepetition := proc(n::posint, r::posint)
>  local repeatlist, i;
>      repeatlist := [ seq( i $ r, i=1..n) ];
>      combinat[choose](repeatlist, r);
>  end:
>
>  RCombRepetition(3,2);
```

$$[[1, 1], [1, 2], [1, 3], [2, 2], [2, 3], [3, 3]]$$

```
>  RCombRepetition(4,3);
```

$$[[1, 1, 1], [1, 1, 2], [1, 1, 3], [1, 1, 4], [1, 2, 2], [1, 2, 3], [1, 2, 4], [1, 3, 3],$$
$$[1, 3, 4], [1, 4, 4], [2, 2, 2], [2, 2, 3], [2, 2, 4], [2, 3, 3], [2, 3, 4], [2, 4, 4],$$
$$[3, 3, 3], [3, 3, 4], [3, 4, 4], [4, 4, 4]]$$

(Notes about the procedure: The i $ r means repeat i r times.

```
>     1 $ 3;
```

$$1, 1, 1$$

```
>  happy $ 4;
```

$$happy, happy, happy, happy$$

Also, we need to use a list instead of a set, since Maple automatically removes repeated elements in a set and we would lose all the repetitions.)

```
>  happylist := [ happy $ 4];
```

$$happylist := [happy, happy, happy, happy]$$

```
>  happyset := { happy $ 4 };
```

$$happyset := \{happy\}$$

1. Find the number of possible outcomes in a two-team playoff when the winner is the first team to win 5 out of 9, 6 out of 11, 7 out of 13 and 8 out of 15 games.

Solution: Our solution will use the Maple procedure called **permute** in order to compute the total number of ways a tournament of games can be played out. We will begin by constructing two lists that keep track of how each of the two teams may win. We will assign the two possibilities of team 1 winning the tournament without any losses, and team 2 winning the tournament without any losses. At each iteration of the algorithm's main loop, we will compute the possible permutations of games to be played, noting that the order of wins is important to us. After these permutations have been calculated, we will increase the number of games that the tournament lasts (i.e. allow the eventual tournament losing team to win one additional game).

This is equivalent to using a tree diagram to compute the possible results. The outer (while) loop corresponds to the level of vertices in the tree, and the inner (for) loop iterates over all the games at that level.

The Maple implementation of this description is shown below.

```
>   Tournaments:=proc(games::integer)
>       local i, one_wins, two_wins, Temp, S;
>       # Initialize a list to guarantee that team 1 wins
>       one_wins:=[seq(1, i=1..ceil(games/2))];
>       # Initialize a list to guarantee that team 2 wins
>       two_wins:=[seq(2, i=1..ceil(games/2))];
>       S:={};
>       # Loop until we have all the games of the series used
>       while nops(one_wins) <= games do
>           # Calculate the possible outcomes that complete in
>           # exactly games games
>           Temp:=permute(one_wins);
>           for i from 1 to nops(Temp) do
>               # Ensure that we really needed all the games
>               # (i.e. the last game of the series was won by team 1)
>               if Temp[i][nops(one_wins)] = 1 then
>                   S:=S union {Temp[i]}
>               fi;
>           od;
>           # Calculate the possible outcomes that complete in
>           # exactly games games
>           Temp:=permute(two_wins);
>           for i from 1 to nops(Temp) do
>               # Ensure that we really needed all the games
>               # (i.e. the last game of the series was won by team 1)
>               if Temp[i][nops(two_wins)] = 2 then
>                   S:=S union {Temp[i]}
>               fi;
>           od;
>           # Increment the number of games, so that the tournament
>           # winning team loses one more game.
>           one_wins:=[op(one_wins), 2];
>           two_wins:=[op(two_wins), 1];
>       od;
>       S;
>   end:
```

We now try this newly created procedure on tournaments that are best 3-of-5 and best 4-of-7 in number of games.

```
>  Tournaments(5);
```
$$\{[2,2,1,1,2],[2,1,2,1,2],[2,1,1,2,2],[1,2,2,1,2],[1,2,1,2,2],$$
$$[1,1,2,2,2],[2,1,1,1],[1,2,1,1],[1,1,2,1],[1,2,2,2],[2,1,2,2],$$
$$[2,2,1,2],[1,2,2,1,1],[1,2,1,2,1],[1,1,2,2,1],[2,1,1,2,1],$$
$$[2,1,2,1,1],[2,2,1,1,1],[2,2,2],[1,1,1]\}$$

```
>  nops(");
```
$$20$$

```
>  nops(Tournaments(7));
```
$$70$$

The reader is left to explore the remaining cases, and conjecture a formula in the general case.

3. We want to look at the binomial coefficients $C(2n, n)$. Specifically, for many examples, we want to determine if $C(2n, n)$ is divisible by the square of a prime, and whether the largest exponent in the prime factorization grows without bound as n grows.

Solution: We will first try one example, to see what exactly we wish to do, and then write a program.

```
>  c := binomial(6,3);
```
$$c := 20$$

We use the function `ifactors` (the "i" stands for *integer*) to factor c. This function is one of several in Maple that must be readlib-defined before we can use it. This means that we ask Maple to find the function in its library, and load it into the current session.

```
>  readlib(ifactors):
>  ifacts := ifactors(c);
```
$$ifacts := [1, [[2, 2], [5, 1]]]$$

The help page for `ifactors` explains what this result means. It says that $20 = 1 \cdot 2^2 \cdot 5^1$. We are interested in the exponents of the primes. First, we take the second element of the list, to get the list of primes and exponents.

```
>  facts := ifacts[2];
```
$$facts := [[2, 2], [5, 1]]$$

This gives us a list of lists, where the first element in each list is the prime factor, and the second is the multiplicity (the number of times that factor appears) of that prime. So we want to traverse this list and take the second element of each sublist.

```
> powers := seq(x[2],x=facts);
```
$$powers := 2, 1$$

Then we use the max function to find the largest exponent.

```
> max(powers);
```
$$2$$

If the largest example is greater than 1, then $C(2n, n)$ is divisible by the square of a prime. In this case, the largest example 2 is, in fact, greater than 1, and $C(6, 3)$ is indeed divisible by 5^2.

Combining these steps, we now write a program that given n, returns the largest exponent in the factorization of $C(2n, n)$.

```
> LargestExpon := proc(n)
> local c, ifacts, x;
>       c := binomial(2*n,n);
>       ifacts := ifactors(c);
>       max(seq(x[2],x=ifacts[2]));
> end:
> LargestExpon(6);
```
$$2$$

Now we write another routine that will calculate the largest exponent for many values of n, and store the results in a table.

```
> Manyn := proc(maxn)
> local results, i;
>    for i to maxn do
>         results[i] := LargestExpon(i);
>         if results[i] = 1 then
>             printf('Hurray! A counterexample! %d\n',
>                   i);
>         fi;
>    od;
>    eval(results);
> end:
```

Run the program and see what happens.

```
> Manyn(10):

Hurray! A counterexample! 1
Hurray! A counterexample! 2
Hurray! A counterexample! 4
```

It seems that 1, 2 and 4 are values of n such that $C(2n, n)$ is not divisible by the square of a prime.

```
>  binomial(8,4);
```

$$70$$

```
>  ifactors(");
```

$$[1, [[2, 1], [5, 1], [7, 1]]]$$

Now let the program run for much longer, and see if we can find anymore.

```
>  vals := Manyn(200):
```

```
Hurray! A counterexample! 1
Hurray! A counterexample! 2
Hurray! A counterexample! 4
```

```
>
```

Let's look at the growth of the maximum exponent by graphing the results.

```
>  plot([ seq([i,vals[i]],i=1..200)],style=POINT,
>  title='Growth of Largest Exponents');
```

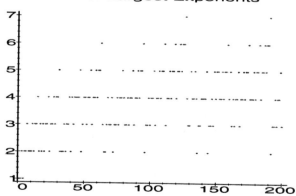

For comparison, try again with even more values of n.

```
>  vals := Manyn(300):
```

```
Hurray! A counterexample! 1
Hurray! A counterexample! 2
Hurray! A counterexample! 4
```

This time, plot with the points joined, to see what difference that makes.

```
> plot([ seq([i,vals[i]],i=1..300)],
> title='Growth of Largest Exponents 2');
```

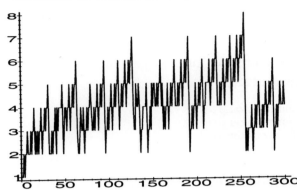

Growth of Largest Exponents 2

It's hard to reach any conclusions from these two graphs, other than there does not appear to be a bound to the size. The time of calculation is becoming long, but we can still look at some largest examples.

```
> LargestExpon(500);
```
$$6$$

```
> LargestExpon(1001);
```
$$7$$

```
> LargestExpon(1005);
```
$$8$$

```
> LargestExpon(1007);
```
$$9$$

```
> LargestExpon(1009);
```
$$7$$

6. Estimate the probability that two integers chosen at random are relatively prime by testing a large number of randomly selected pairs of integers. Look up the theorem that gives this probability and compare your results with the correct probability.

Solution: To solve this problem, three things must be done.

(i) Devise a method for generating pairs of random integers.

(ii) Produce a large number of these pairs, testing whether they are relatively prime, and noting the probability estimate based on this sample.

(iii) Look up the theorem mentioned in the question.

Naturally, we'll leave part (iii) entirely to the reader.

A simple approach is to use the Maple procedure rand to generate a list of random integers. Then, having generated such a list we can test the coprimality of its members in pairs using the Maple procedure igcd in a second loop. We implement these two loops in a new Maple procedure called RandPairs:

```
>   RandPairs := proc(list_size::integer)
>      local i, tmp, randnums, count;
>      randnums := NULL;
>      # Generate list of random integers
>      for i from 1 to list_size do
>        tmp := rand();
>        randnums := randnums, tmp();
>      od;
>      randnums := [randnums];
>      # Count the number of pairs that are coprime
>      count := 0;
>      for i from 1 by 2 to list_size-1 do
>        if igcd(randnums[i], randnums[i + 1]) = 1 then
>           count := count + 1;
>        fi;
>      od;
>      count;
>   end:
```

We can now execute this procedure on 100 pairs of integers, as follows:

```
>   RandPairs(200);
```
$$64$$

Then, we can determine the percentage of coprime pairs by using this result;

```
>   evalf(RandPairs(200)/100);
```
$$.6400000000$$

Note that repeating the identical computation may very well lead to a somewhat different result since the list of integers we used was generated "randomly". You should try this with a much larger sample size, say 1000 pairs of integers.

7. Determine the number of people needed to ensure that the probability at least two of them have the same day of the year as their birthday is at least 70 percent, at least 80 percent, at least 90 percent, at least 95 percent, at least 98 percent, and at least 99 percent.

Solution: Given that we know the formula for the probability of two people having the same birthday on the same day, we can use Maple to loop over a range of possible numbers of people, until we reach a probability greater than the desired probability.

If we consider the probability that no two people have the same birthday as p, we can determine the probability that at least two people are born on the same day of the year as $1 - p$. To determine what p is, we note that if we have k people, the first person has probability 1 of having the same birthday as himself. The second person has 364 other days out of 365 to choose from so that they do not have a birthday on the same day as the first person. Similarly for the person $3, 4, \ldots, k$, where the kth person has $365 - k$ choices. Taking the product of these probabilities, we conclude that $p = P(365, k)/365^k$, which allows us to easily compute $1 - p$.

We now represent and combine this information in a Maple procedure called Birthdays;

```
> Birthdays := proc(percentage::float)
>    local num_people, cur_prob;
>    # Initialize
>    cur_prob := 0; num_people:=0;
>    # Loop until enough people
>    while cur_prob < percentage do
>      num_people := num_people + 1;
>      cur_prob := 1
>         -(numbperm(365,num_people) / 365^num_people);
>    od;
>    RETURN(num_people);
> end:
```

This procedure returns the number of people required to attain the given probability that two have the same birthday. We now execute our procedure on a few test cases, for probabilities of $0.70, 0.80$ and 0.90;

```
> Birthdays(.70);
```
$$30$$

```
> Birthdays(.90);
```
$$41$$

4.7   Exercises/Projects

Exercise 1. By using Maple to generate many rows of Pascal's triangle, see if you can formulate any conjectures involving identities satisfied by the binomial coefficients $C(n, k)$.

Exercise 2. Use maple to determine how many different strings can be made from the word "PAPARRAZZI" when all the letters are used; when any number of letters are used; when all the letters are used and the string begins and ends with the letter "Z"; when all the letters are used and the three "A"'s are consecutive.

Exercise 3. Use the Pigeon Hole Principle to design, and then implement, a Maple procedure that finds a maximal increasing subsequence of a given sequence of numbers. (See Page 245, *et seq* in your text.)

Exercise 4. Suppose that a certain Mathematics Department has m male faculty and f female faculty. Write a Maple procedure to find all committees with $2k$ members in which both sexes are represented equally.

Exercise 5. Use Maple to prove the identity

$$C(n + 1, k) = (n + 1)C(n, k - 1)/k.$$

, for positive integers n and k with $k \leq n$.

Exercise 6. Use Maple to prove Pascal's identity: $C(n + 1, k) = C(n, k - 1) + C(n, k)$, for all positive integers n and k with $n \geq k$.

Exercise 7. Use Maple to determine the integer k such that the chances of picking six numbers correctly in a lottery from the first k positive integers is less than

(a) 1 in 100 million,

(b) 1 in a billion (10^9),

(c) 1 in 10 billion,

(d) 1 in 100 billion, and

(e) 1 in a trillion (10^{12}).

Exercise 8. Use Maple to count and list all solutions to the equation

$$x_1 + x_2 + x_3 + x_4 = 25,$$

where x_1, x_2, x_3 and x_4 are non-negative integers.

Exercise 9. Generate a large triangle of Stirling numbers and look for patterns that suggest identities among the Stirling numbers. (A small triangle was shown in Section 4.2.) Can you make any conjectures about the relationship between Stirling numbers and the binomial coefficients?

Exercise 10. Write a Maple function that takes as input three positive integers n, k and i, and returns the ith multinomial, in lexicographic order, of the polynomial $(x_1 + x_2 + \cdots + x_k)^n$. Write its inverse; that is, given a multinomial, the inverse should return its index (position) in the sorted polynomial.

Exercise 11. Write a Maple program to compute the Cantor expansion of an integer. (See Page 298 of the textbook.)

Exercise 12. Implement, in Maple, the algorithm for generating the set of all permutations of the first n integers, using the bijection from the collection of all permutations of the set $\{1, 2, \ldots, n\}$ to the set $\{1, 2, \ldots, n!\}$ described prior to Exercise 10 on Page 298 of the textbook.

Exercise 13. Write a Maple procedure to generate random permutations as described in Exercise 14 on Page 298 of the textbook.

5 Counting

In this chapter we will describe how to use Maple to work with three important topics in counting: recurrence relations, inclusion-exclusion, and generating functions. We begin by describing how Maple can be used to solve recurrence relations, including the recurrence relation for the sequence of Fibonacci numbers. We then show how to solve the Tower of Hanoi puzzle and we find the number of moves required for n disks. We describe how Maple can be used to solve linear homogeneous recurrence relations with constant coefficients, as well as the related inhomogeneous recurrence relations. After describing how to solve these special types of recurrence relations with Maple, we show how to use Maple's general recurrence solver. We illustrate the use of this general solver by demonstrating how to use it to solve divide and conquer recurrence relations. After studying recurrence relations, we show how to use Maple to help solve problems using the principle of inclusion and exclusion. Finally, we discuss how Maple can be used to work with generating functions, a topic covered in Appendix 3 in the text.

5.1 Recurrence Relations

A recurrence relation describes a relationship that one member of a sequence $\{a_n\}$ of values has to other member of the sequence which precede it. For example, the famous Fibonacci sequence $\{F_n\}$ satisfies the recurrence relation

$$F_n = F_{n-1} + F_{n-2}.$$

Together with the initial conditions $F_1 = 1$ and $F_2 = 1$, this relation is sufficient to define the entire sequence $\{F_n\}$.

In general, we can think of a recurrence relation as a relation of the form

$$r_n = f(r_{n-1}, r_{n-2}, \ldots, r_{n-k}),$$

in which each term r_n of the sequence depends on some number k of the terms which precede it in the sequence. For example, for the Fibonacci sequence, the function f is $f(x, y) = x + y$.

To understand how we can work with recurrence relations in Maple, we have to stop for a moment and realize that a sequence $\{r_n\}$ of "values" (numbers, matrices, circles, functions, etc.) is just a *function* whose domain happens to be the set of (usually positive) integers. If we want to take this point of view (and we do!), then the nth term r_n of a sequence $\{r_n\}$ would be more conventionally written as $r(n)$, and we would refer to the *function* r. In this way, we can think of the sequence $\{r_n\}$ as one way of representing a function r whose domain is the set of positive integers, and whose value at the integer n is just $r_n = r(n)$. This really just amounts to a change in notation; there is nothing more to it.

Once this change in notation has been made, it is then easy to see how to represent a recurrence relation as a Maple procedure taking integer arguments.

In chapter 3 (see page 109) we discovered how to efficiently represent the *Fibonacci* sequence by the procedure

```
>  Fibonacci := proc(n::posint) option remember;
>     if   n = 1   or n = 2 then RETURN( 1 ); fi;
>     Fibonacci2(n-1) + Fibonacci2(n-2);
>  end:
```

Recall that the first line of this procedure instructs Maple to "remember"[1] whatever values of the procedure have already been calculated in the current session.

Sometimes, in spite of our best efforts, a recursive implementation of an algorithm may be too costly simply due to its very nature. A recursive implementation can be avoided if we can find an explicit formula for the general term of the recurrence. The process of finding such an explicit formula is referred to as "solving" the recurrence. In the next section, we shall see how to use Maple to do this for certain kinds of recurrence relations.

Towers of Hanoi Problem

The famous puzzle known as the "Towers of Hanoi Problem" is discussed in the text, where the recurrence relation

$$H_n = 2H_{n-1} + 1, \quad H_1 = 1$$

is derived, in which H_n denotes the number of moves required to solve the puzzle for n disks. As discussed in the text, this has the solution

$$H_n = 2^n - 1.$$

Later, we shall see how to use Maple to derive this result quite simply.

---

[1] You can cause Maple to "forget" values by using the Maple command `forget`.

Besides solving for the number of moves required to solve the Towers of Hanoi Problem for n disks, we can illustrate the solution by writing a Maple program to compute the moves needed to solve the Towers of Hanoi Problem, and describing them for us. We'll write a small program consisting of three Maple procedures: the main program Hanoi, a utility routine PrintMove, and the recursive engine of the program TransferDisk, which does most of the work.

The easiest part to write is the function PrintMove, which merely displays for us the move to make at a given step.

```
> PrintMove := proc(src::string, dest::string)
>    printf('Move disk from peg %s to peg %s\n',
>          src, dest);
> end:
```

Here, we just call the Maple library procedure printf, which may be used for formatted output. The function printf has a complex calling syntax; refer to the online help for details and further information. (Note: If you are familiar with the printf function in C, then you will find that Maple's version of printf is quite similar. In this case, the symbols %s above are replaced by the string values of the second and third arguments, respectively.)

Next, the recursive procedure TransferDisk does most of the work for us. This function models the idea of transferring a disk from one peg to another. But, because it is recursive, we need to supply to it, as an argument, the total number of disks to be handled in each call.

```
> TransferDisk := proc(src::string, via::string,
>    dest::string, ndisks::posint)
>    if ndisks = 1 then
>         PrintMove(src, dest);
>    else
>         TransferDisk(src, via, dest, ndisks -1);
>         PrintMove(src, dest);
>         TransferDisk(via, dest, src, ndisks -1);
>    fi;
> end:
```

Finally, we package it in a top level procedure, Hanoi, thereby providing an interface to the recursive engine.

```
> Hanoi := proc(ndisks::posint)
>    if ndisks < 1 then
>         printf('What's wrong with this picture?\n');
>    else
>         TransferDisk('A', 'B', 'C', ndisks);
>    fi;
> end:
```

Our Hanoi program can exhibit a specific solution to the Towers of Hanoi Problem for any number ndisks of disks.

```
>  Hanoi(2);
```

```
Move disk from peg A to peg C
Move disk from peg A to peg C
Move disk from peg B to peg A
```

```
>  Hanoi(3);
```

```
Move disk from peg A to peg C
Move disk from peg A to peg C
Move disk from peg B to peg A
Move disk from peg A to peg C
Move disk from peg B to peg A
Move disk from peg B to peg A
Move disk from peg C to peg B
```

Try experimenting with different values of ndisks to get a feel for how large the problem becomes for even moderately large values of ndisks.

5.2 Solving Recurrences with Maple

Now that we know how to implement recurrence relations in Maple, and we have worked with them a little, we will see how to use Maple to solve certain kinds of recurrence relations.

Maple has a very powerful recurrence solver rsolve that we shall discuss later. Its use, however, can obscure some of the important ideas that are involved. Therefore, we shall first use some of Maple's more primitive facilities to solve certain kinds of recurrence relations one step at a time.

Given a recursively defined sequence $\{r_n\}$, what we would like is to find some kind of "formula", involving only the index n (and, perhaps, other fixed constants and known functions) which does *not* depend on knowing the value of r_k, for any index k.

To begin with, we shall consider recurrence relations that are *linear, homogeneous*, and which have *constant coefficients*; that is, they have the form

$$r_n = a_1 r_{n-1} + a_2 r_{n-2} + \cdots + a_k r_{n-k},$$

where $a_1, a_2, \ldots, a_k$ are real constants and a_k is nonzero. Recall that the integer k is called the *degree* of this recurrence relation. To have a unique solution, at least k initial conditions must be specified.

The general method for solving such a recurrence relation involves finding the roots of its characteristic polynomial

$$x^k - a_1 x^{k-1} - a_2 x^{k-2} - \cdots - a_{k-1} x - a_k.$$

When this polynomial has distinct roots, all solutions are linear combinations of the nth powers of these roots. When there are repeated roots, the situation is a little more complicated, as we shall see.

To begin with, let's consider a linear homogeneous recurrence relation with constant coefficients of degree two:

$$r_n = 2r_{n-1} + 3r_{n-2},$$

subject to the initial conditions

$$r_1 = 4 \qquad \text{and} \qquad r_2 = 2.$$

Then its characteristic equation is

$$x^2 - 2x - 3.$$

To solve the recurrence relation, we must solve for the roots of this equation. Using Maple makes this very easy; we use the `solve` function in Maple to do this.

```
>   solve(x^2 - 2*x - 3 = 0, x);
                              -1, 3
```

The syntax tells the `solve` function that we want the values of the variable x which satisfy the quadratic equation

```
x^2 - 2 * x - 3 = 0.
```

Now that Maple has told us that the solutions are $x = 3$ and $x = -1$, we can write down the form of the solution to the recurrence as

$$r_n = \alpha 3^n + \beta(-1)^n,$$

where α and β are constants that we have yet to determine. We can use Maple to determine the constants α and β. Since the initial conditions are $r_1 = 4$ and $r_2 = 2$, we know that our recurrence relation must satisfy the following two equations.

$$3\alpha - \beta = 4 \qquad\qquad (5.1)$$
$$3^2\alpha + \beta = 2 \qquad\qquad (5.2)$$

To find the solutions of this system of linear equations, we use Maple's solve facility:

```
> solve( {3 * alpha - beta = 4, 9 * alpha + beta = 2},
> {alpha, beta});
```

$$\{\alpha = \frac{1}{2}, \beta = \frac{-5}{2}\}$$

This time, we are telling Maple to solve the *set* of equations as indicated, so we use Maple's notation for sets. Likewise, the solutions form a *set* $\{\alpha, \beta\}$, so we must tell Maple to solve for a set of variables in this case. Thus, the complete solution to our recurrence is

$$r_n = \frac{1}{2} \cdot 3^n + \frac{-5}{2} \cdot (-1)^n.$$

This formula allow us to write a very efficient Maple function for finding the terms of the sequence $\{r_n\}$, which is obviously much more efficient than a recursive procedure.

```
> r := proc(n) ((3^n)/2 - (5*((-1)^n)))/2  end:
```

Let's try another example. We'll solve the recurrence relation

$$r_n = \frac{-5}{3} r_{n-1} + \frac{2}{3} r_{n-2}.$$

with the initial conditions

$$r_1 = \frac{1}{2} \qquad \text{and} \qquad r_2 = 4.$$

To do this, we ask Maple to solve the characteristic polynomial of the recurrence relation, and then to solve the system of linear equations which results from use of the initial conditions. Observe that this method works because this recurrence relation is linear, homogeneous and has constant coefficients.

```
> evals := solve(x^2 + (5/3) * x - 2/3 = 0, x);
```

$$evals := -2, \frac{1}{3}$$

```
> solve({
>    alpha * evals[1] + beta * evals[2]   =  1/2,
>    alpha * evals[1]^2 + beta * evals[2]^2 = 4
> },
> {alpha, beta});
```

$$\{\alpha = \frac{23}{28}, \beta = \frac{45}{7}\}$$

(This time, we have named the sequence of solutions to the characteristic equation `evals` so that we can more easily use them in the calls to `solve`.)

Thus we see that the solution to the recurrence relation is

$$r_n = \frac{45}{7}(\frac{1}{3})^n + \frac{23}{28}(-2)^n.$$

We can derive the explicit formula for the Fibonacci sequence this way as well. The characteristic polynomial for the Fibonacci sequence is

$$x^2 - x - 1.$$

Solving for its roots yields

```
>   evals := solve(x^2 - x - 1, x);
```

$$evals := \frac{1}{2}\sqrt{5} + \frac{1}{2}, \frac{1}{2} - \frac{1}{2}\sqrt{5}$$

We find the coefficients α and β in the formula for the nth Fibonacci number by using the initial conditions

```
>   sol := solve({
>     alpha * evals[1] + beta * evals[2] = 1,
>     alpha * evals[1]^2 + beta * evals[2]^2 = 1
>   },
>   {alpha, beta});
```

$$sol := \{\alpha = \frac{1}{5}\sqrt{5}, \beta = -\frac{1}{5}\sqrt{5}\}$$

The formula for the nth Fibonacci number is just

```
>   rn := alpha*evals[1]^n + beta*evals[2]^n;
```

$$rn := \alpha(\frac{1}{2}\sqrt{5} + \frac{1}{2})^n + \beta(\frac{1}{2} - \frac{1}{2}\sqrt{5})^n$$

You can use `sol` and the `subs` command to insert the values for `alpha` and `beta` into this formula `sol`. The result is

```
>   rn := subs(sol,rn);
```

$$rn := \frac{1}{5}\sqrt{5}(\frac{1}{2}\sqrt{5} + \frac{1}{2})^n - \frac{1}{5}\sqrt{5}(\frac{1}{2} - \frac{1}{2}\sqrt{5})^n$$

It is no accident that the equations returned by this form of the `solve` command are in exactly the right form to be used in the `subs` command.

If we are to use such a formula to repeatedly to compute values, then we should use it to define a function. You can "type in" a new function definition, but a much more convenient way is to use the `unapply` command[2]

---

[2] The `unapply` command is named this way because it is in effect the reverse of *applying* a function.

which takes as its arguments an expression and the variables that are to be used to define the function. The resulting procedure is

> `Fibonacci2 := unapply( rn , n );`

$$Fibonacci2 := n \rightarrow \frac{1}{5}\sqrt{5}\left(\frac{1}{2}\sqrt{5}+\frac{1}{2}\right)^n - \frac{1}{5}\sqrt{5}\left(\frac{1}{2}-\frac{1}{2}\sqrt{5}\right)^n$$

The procedure `Fibonacci2` is even more efficient than the even the optimized recursive procedure `Fibonacci`. To see this, we record the accumulated time for computing the first 100 Fibonacci numbers.

> `st:=time(): for i to 100 do Fibonacci(i): od: time() - st;`

.260

> `st:=time(): for i to 100 do Fibonacci2(i): od: time() - st;`

.070

Recall that the naively coded procedure `Fibonacci` is so inefficient that it cannot be used to compute the 100th Fibonacci number (see page 109.

A General Linear Homogeneous Recurrence Relation with Constant Coefficients Solver

Now let's generalize what we have been doing and write a Maple procedure to solve a general degree two linear, homogeneous recurrence relation with constant coefficients, provided that the roots of the characteristic polynomial of the recurrence relation are distinct. We'll write a procedure `RecSol2` which solves the recurrence

$$r_n = ar_{n-1} + br_{n-2}$$

subject to the initial conditions

$$r_1 = u \qquad \text{and} \qquad r_2 = v.$$

and then returns a procedure that can be used to compute terms of the sequence.

For the moment, assume that the characteristic polynomial $x^2 - ax - b$ has two distinct roots. Then, all our procedure need do is to repeat the steps we did manually in our earlier example.

```
>   RecSol2 := proc(a, b, u, v)
>      local evals, S, alpha, beta, ans , n;
>      #  u solve the characteristic equation
>      evals := solve(x^2 - a * x - b = 0, x);
>      #  next solve the system of linear equations
>      S := solve({alpha * evals[1] + beta * evals[2] = u,
```

```
>            alpha * evals[1]^2 + beta * evals[2]^2 = v},
>            {alpha,beta});
>    ans := subs(S,alpha*evals[1]^n + beta*evals[2]^n);
>    RETURN( unapply( ans , n ) );
>  end:
```

To see how it works, we'll try it on some test cases. To construct a function for computing the Fibonacci sequence, invoke our new procedure as:

```
>   f := RecSol2(1,1,1,1,5);
```

$$f := n \to \frac{1}{5}\sqrt{5}\,(\frac{1}{2}\sqrt{5}+\frac{1}{2})^n - \frac{1}{5}\sqrt{5}\,(\frac{1}{2}-\frac{1}{2}\sqrt{5})^n$$

The resulting procedure can be used to compute the general term of the Fibonacci sequence.

```
>   f(n);
```

$$\frac{1}{5}\sqrt{5}\,(\frac{1}{2}\sqrt{5}+\frac{1}{2})^n - \frac{1}{5}\sqrt{5}\,(\frac{1}{2}-\frac{1}{2}\sqrt{5})^n$$

Likewise, the first five Fibonacci numbers can be computed as follows.

```
>   seq(simplify(f(n)), n = 1..10);
```

$$1, 1, 2, 3, 5, 8, 13, 21, 34, 55$$

We now present a solver that can handle the case of repeated roots.

Before we look at the new version of `RecSol2`, let's look at an example involving a recurrence relation with a double eigenvalue (root of its characteristic polynomial). The recurrence relation

$$r_n = 4r_{n-1} - 4r_{n-2}$$

has the characteristic equation

```
>   char_eqn := x^2 - 4 * x + 4 = 0;
```

$$char\_eqn := x^2 - 4\,x + 4 = 0$$

with eigenvalues

```
>   evals := [solve(char_eqn, x)];
```

$$evals := [2, 2]$$

In general, to test for a repeated eigenvalue, which is the case for this example, we just test whether

```
>   evalb(evals[1] = evals[2]);
```

$$true$$

(Note: We do not require the use of `evalb` in a conditional statement since expressions are automatically evaluated as booleans there.) If we call the double root (2 in this case) λ, then the recurrence relation has the explicit solution

$$r_n = \alpha \lambda^n + n\beta\lambda^n,$$

for all positive integers n, and for some constants α and β. Assuming initial conditions of $r_1 = 1$ and $r_2 = 4$, the set S of equations to solve is

```
>   S := {alpha * evals[1] + beta * evals[2] = 1,
>          alpha * evals[1]^2 + 2* beta * evals[2]^2 = 4};
```

$$S := \{2\alpha + 2\beta = 1, 4\alpha + 8\beta = 4\}$$

As before, to get the solutions, we type

```
>   rsols := solve(S, {alpha, beta});
```

$$rsols := \{\beta = \frac{1}{2}, \alpha = 0\}$$

It is at this point that the difference with the case of distinct roots appears. The nth term of the sequence, when there is a double eigenvalue, is given by

```
>   subs(rsols , alpha * evals[1]^n + n * beta * evals[1]^n );
```

$$\frac{1}{2} n\, 2^n$$

The steps carried out in this example are really quite general. A general procedure for solving a two term recurrence of the form $r(n) = ar(n-1) + br(n-2)$, with initial values $r(1) = u$ and $r(2) = v$ is:

```
>   RecSolver2 := proc(a,b,u,v)
>     local ans, evals, S, alpha, beta, rsols, n;
>     # solve the characteristic equation
>     evals := solve(x^2 - a * x - b = 0, x);
>     # solve the system of linear equations
>     S := {alpha * evals[1] + beta * evals[2] = u,
>          alpha * evals[1]^2 + beta * evals[2]^2 = v};
>     rsols := solve(S, {alpha, beta});
>     if evals[1] = evals[2] then # repeated roots
>       ans := subs(rsols,alpha*evals[1]^n + beta*n*evals[1]^n);
>     else
>       ans := subs(rsols,alpha*evals[1]^n + beta*evals[2]^n );
>     fi;
>     RETURN( unapply(ans , n ) );
>   end:
```

This version of our solver firsts test for a repeated root, and then does the appropriate computation, based on the result. It is invoked in the same way that `RecSol2` is.

```
>   g := RecSolver2(4,-3,1,2);
```

$$g := n \to \frac{1}{2} + \frac{1}{6} 3^n$$

```
>   i :='i': seq(simplify(g(i)), i=1..10);
```

$$1, 2, 5, 14, 41, 122, 365, 1094, 3281, 9842$$

This gives the first ten terms of the sequence defined by the recurrence relation $r_n = 4r_{n-1} - 3 * r_{n-2}$, with initial conditions $r_1 = 1$ and $r_2 = 2$.

To solve the recurrence $r_n = -r_{n-1} - r_{n-2}$, with initial conditions $r_1 = 1$ and $r_2 = 2$, we use the The solution and the first 10 terms of this sequence are

```
>   h := RecSolver2(-1,-1,1,2);
```

$$h := n \to (-\frac{3}{2} + \frac{1}{6} I \sqrt{3}) (-\frac{1}{2} + \frac{1}{2} I \sqrt{3})^n + (-\frac{3}{2} - \frac{1}{6} I \sqrt{3}) (-\frac{1}{2} - \frac{1}{2} I \sqrt{3})^n$$

```
>   i := 'i': seq(simplify(h(i)),i=1..10);
```

$$1, 2, -3, 1, 2, -3, 1, 2, -3, 1$$

Notice the pattern that emerges if we replace the initial conditions $r_1 = 1$ and $r_2 = 2$ with symbolic constants.

```
>   k := RecSolver2(-1, -1, lambda, mu);
```

$$k := n \to (-\frac{1}{2} \lambda - \frac{1}{2} \mu + \frac{1}{6} I \sqrt{3} \mu - \frac{1}{6} I \lambda \sqrt{3}) (-\frac{1}{2} + \frac{1}{2} I \sqrt{3})^n$$
$$+ (-\frac{1}{2} \lambda - \frac{1}{2} \mu + \frac{1}{6} I \lambda \sqrt{3} - \frac{1}{6} I \sqrt{3} \mu) (-\frac{1}{2} - \frac{1}{2} I \sqrt{3})^n$$

```
>   i := 'i': seq(simplify(k(i)),i=1..10);
```

$$\lambda, \mu, -\lambda - \mu, \lambda, \mu, -\lambda - \mu, \lambda, \mu, -\lambda - \mu, \lambda$$

Inhomogeneous Recurrence Relations

We have, so far, been discussing *homogeneous* linear recurrence relations with constant coefficients. However, the techniques used in solving them may be extended to provide solutions to *inhomogeneous* recurrences of this type. These are recurrence relations of the form

$$\alpha_n r_n + \alpha_{n-1} r_{n-1} + \cdots + \alpha_{n-k} r_{n-k} = c_n$$

where $\alpha_n, \alpha_{n-1}, \ldots, \alpha_{n-k}$ and c_n are constants. The only new wrinkle is that, here, the c_n need not be zero. Put another way, an equation of this form in which every c_n is zero is a homogeneous one, so the homogeneous relations are just a special case of this more general type. To solve the more general recurrence, we need to do two things:

1. Find on specific solution to the inhomogeneous recurrence

2. Solve the corresponding homogeneous recurrence.

The corresponding homogeneous recurrence is just the one obtained by replacing the sequence $\{c_n\}$ by the zero sequence:

$$\alpha_n r_n + \alpha_{n-1} r_{n-1} + \cdots + \alpha_{n-k} r_{n-k} = 0.$$

So, we already know how to do the second step.

The first step is more difficult, but with the help of Maple, it is rendered manageable.

```
>  rsolve({r(0) = 0, r(n) = 3* r(n-1) + 3^n}, r(n));
```
$$(n+1)\,3^n - 3^n$$

```
>  normal(",expanded);
```
$$3^n\,n$$

This tells us that $r_n = n3^n$ is one solution to the recurrence relation $r_n = 3r_{n-1} + 3^n$. Now, all solutions are obtained by adding this one solution to the set of solutions of the corresponding homogeneous recurrence.

```
>  rsolve(r(n) = 3 * r(n-1), r(n));
```
$$r(0)\,3^n$$

```
>  " + n * 3^n;
```
$$r(0)\,3^n + n\,3^n$$

If we have an initial value for r_0, then we have a complete solution.

Now let us solve the Tower of Hanoi recurrence

$$H_n = 2H_{n-1} + 1$$

which gives the number of moves necessary to solve to Towers of Hanoi puzzle with n disks. Remember that $H_1 = 1$. (See Page 312 of the text.) The associated homogeneous recurrence relation is

$$h_n = 2h_{n-1}$$

with characteristic polynomial

$$x - 2.$$

The only root of this is 2, so all solutions of the homogeneous recurrence relation have the form

$$h_n = \alpha 2^{n-1}$$

for some constant α. (The power of 2 is $n-1$, rather than n, because the recurrence starts at 1 instead of 0.) Solutions for H are obtained from the solutions for h by adding a particular solution for H. Now, H has the constant solution $H_n = -1$, for all n, so all solutions for H are of the form

$$H_n = \alpha 2^n - 1.$$

Using the initial condition

$$H_1 = 1$$

we can solve for α as follows.

```
>   solve(alpha * 2^1 - 1 = 1, alpha);
                        1
```

Thus, the solution to the Towers of Hanoi recurrence is $H_n = 2^{n-1} - 1$.

Maple's Recurrence Solver

Now that we have seen how it is possible to use Maple to implement an algorithm to solve simple recurrence relations, it is time to introduce Maple's own facility for working with recurrence relations. We have already seen the Maple command `solve` for working with polynomial equations and systems of equations. Similarly, there is a Maple command `rsolve`, which is specially engineered for dealing with *recurrence* relations. It is a sophisticated version of our `RecSol2` procedure, which can deal with recurrence relations of arbitrary degree, and can handle repeated roots, as well as nonlinear recurrence relations. To use `rsolve`, you need to tell it what the recurrence relation is, and some initial conditions. You must also specify the name of the recursive function to solve for. For instance, to solve the Fibonacci recurrence, you can type

```
>   rsolve( {f(n) = f(n-1) + f(n-2), f(0) = 0, f(1) = 1}, f(n));
```

$$\frac{(1 - \frac{1}{5}\sqrt{5})\,(\frac{2}{\sqrt{5}-1})^n}{\sqrt{5} - 1} + \frac{(-1 - \frac{1}{5}\sqrt{5})\,(-\frac{2}{\sqrt{5}+1})^n}{\sqrt{5} + 1}$$

```
>   normal(",expanded);
```

$$-\frac{1}{5}\frac{\sqrt{5}\,(-2)^n}{(\sqrt{5} + 1)^n} + \frac{1}{5}\frac{\sqrt{5}\,2^n}{(\sqrt{5} - 1)^n}$$

It is not actually necessary to specify the initial conditions for a recurrence relation. If they are not present, Maple will still solve the equation, inserting symbolic constants (here, g(0) and g(1)) in place of numeric ones, as the following example illustrates.

```
>  rsolve(g(n) = 2*g(n-1) - 6*g(n-2), g(n));
```

$$\frac{\left(\frac{1}{10}\, I\, g(1)\, \sqrt{5} - \frac{1}{2}\, g(1) + \frac{2}{5}\, I\, g(0)\, \sqrt{5} + g(0)\right)\left(\frac{6}{1+I\sqrt{5}}\right)^n}{1 + I\sqrt{5}}$$

$$+\; \frac{1}{10}\, \frac{I\sqrt{5}\,(g(1) - I\, g(1)\, \sqrt{5} + 4\, g(0) + 2\, I\, g(0)\, \sqrt{5})\left(-\frac{6}{-1+I\sqrt{5}}\right)^n}{-1 + I\sqrt{5}}$$

We see, in this formula, that Maple uses the symbol I to denote the imaginary unit $(\sqrt{-1})$.

The function `rsolve` can handle several difference kinds of recurrence relations. In Maple V, Release 4, this list includes:

1. linear recurrence relations with constant coefficients

2. systems of linear recurrence relations with constant coefficients

3. divide and conquer recurrence relations with constant coefficients

4. many first order linear recurrence relations

5. some nonlinear first order recurrence relations

The capabilities of `rsolve`, like other Maple functions, are constantly being enhanced and extended. If you have a later release of Maple you may find that your version of `rsolve` has capabilities beyond those enumerated above. However, `rsolve` is not a panacea; you can easily find recurrence relations that it is incapable of solving.

```
>  rsolve(u(n) = u(n-1)^2 - exp(2*u(n-2)), u(n));
```

When `rsolve` is unable to solve a recurrence relation, is simply returns unevaluated.

It is often the case that a problem, as presented, gives no clue that a solution may be found using recurrences. Let's see how we can use Maple to solve a real problem; that is, one that is not explicitly expressed as one requiring the use of recurrences for its solution. Into how many regions is the plane divided by 1000 lines, assuming that no two of the lines are parallel, and no three are coincident? Such a situation may arise in an attempt to model fissures in the ocean floor, or elsewhere on the surface of the earth.

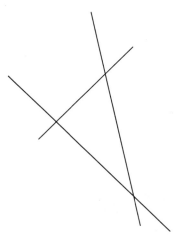

Figure 5.1: Three lines dividing the plane

To start with, we might try to discover the answer for smaller numbers of lines. So, to generalize the problem, we may ask for the number of regions produced by n lines, where n is some positive integer. It is fairly obvious that a single line (corresponding to the case in which $n = 1$) divides the plane into 2 regions. Two lines, if they are not parallel, can easily be seen to divide a plane into 4 regions. (Two distinct parallel lines produce only 3 regions.) If we call the number of regions produced by n lines, no two of which are parallel, and no three of which are coincident r_n, then we have $r_1 = 2$ and $r_2 = 4$. So far, it is beginning to look like $r_n = n^2$. But let's not be hasty. What does the situation look like when $n = 3$? The figure shown here is representative of the situation. In this case, the number r_3 of regions is 7, so the initial guess that r_n is n^2 cannot be right. To find r_4, we must add a fourth line to the diagram. This suggests trying to compute r_4 in terms of r_3, so that we will think of $\{r_n\}$ as a recurrence relation. The figure shows what the situation looks like when a fourth line is added to three existing lines. From the assumptions that no two of the lines can be parallel and that no three pass through a single point, it follows that the new line must intersect each of the existing three lines in exactly one point. This means that the new line passes through exactly three of the regions formed by the original three lines. Each region that it passes through is divided into two regions, so the total number of new regions added by the addition of the fourth line is 3. Thus, $r_4 = r_3 + 3$. Similar arguments for a general configuration of lines reveal that r_n satisfies the recurrence relation

$$r_n = r_{n-1} + (n - 1)$$

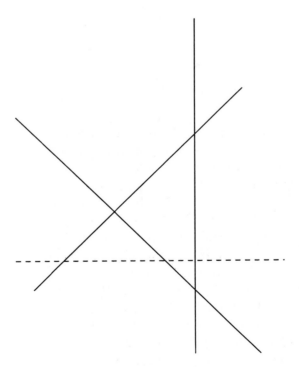

Figure 5.2: Four lines dividing the plane

Furthermore, we have already computed the initial condition $r_1 = 2$. This is enough to solve this recurrence.

```
>  rsolve(
>    {r(n) = r(n-1) + (n-1),
>      r(1) = 2},
>    r(n));
```

$$1 + (n+1)\left(\frac{1}{2}n + 1\right) - 2n$$

```
>  simplify(");
```

$$2 + \frac{1}{2}n^2 - \frac{1}{2}n$$

Divide and Conquer Relations

A very good example of divide and conquer relations is the one provided by the binary search algorithm. Here, we shall consider a practical application of this algorithm in an implementation of a binary search on a sorted list of integers. The algorithm searches for `key` in `ilist`.

```
>  BinSearch := proc(ilist::list(integer), key::integer)
>    local mid, lo, hi;
>    hi := nops(ilist);
>    lo := 0;
>    while hi - lo > 1 do
>      mid := floor((lo + hi) / 2);
>        if key <= ilist[mid] then hi := mid;
>      else lo := mid; fi;
>    od;
>    if ilist[hi] = key then RETURN(hi);
>    else RETURN(false); fi;
>  end:
```

The variable `ilist` is the list of integers to search, and the parameter `key` is the integer to search for. The position in the list is returned if it is found, and the value `false` is returned otherwise. To test `BinSearch`, we use the following little loop with a sample list to search.

```
>  a := [3,5,7,12,34,546,5324,5346753];
```

$$a := [3, 5, 7, 12, 34, 546, 5324, 5346753]$$

```
>  for i in a do
>    if a[BinSearch(a, i)] <> i then
>      print('Socks for President in '96!');
>    fi;
>  od;
```

Unfortunately for Socks, our program worked just fine.

Let us now do the analysis of the algorithm to see how divide and conquer recurrence relations are generated. In general, a divide and conquer type of recurrence relation has the form

$$r_n = ar_{n/k} + b$$

for some constants a, k and b. Now, Maple's `rsolve` routine has absolutely no difficulty handling even the most general type of divide and conquer relation.

```
>   rsolve(r(n) = a * r(n/k) + b, r(n));
```

$$\mathrm{r}(1)\, n^{\left(\frac{\ln(a)}{\ln(k)}\right)} + n^{\left(\frac{\ln(a)}{\ln(k)}\right)} \left(-\frac{b\left(\frac{1}{a}\right)^{\left(\frac{\ln(n)}{\ln(k)}+1\right)} a}{-1+a} + \frac{b}{-1+a} \right)$$

If we know that, say $r_1 = 4$, then we can compute

```
>   subs(r(1) = 4, ");
```

$$4\, n^{\left(\frac{\ln(a)}{\ln(k)}\right)} + n^{\left(\frac{\ln(a)}{\ln(k)}\right)} \left(-\frac{b\left(\frac{1}{a}\right)^{\left(\frac{\ln(n)}{\ln(k)}+1\right)} a}{-1+a} + \frac{b}{-1+a} \right)$$

Each call to the binary search algorithm produces $a = 2$ lists, and each is half the size of the original list ($k = 2$). Therefore, the multiplier and the period in the case of a binary search algorithm are both equal to 2, and so we get

```
>   subs(a = 2, k = 2, ");
```

$$4\, n + n\left(-2\, b\left(\frac{1}{2}\right)^{\left(\frac{\ln(n)}{\ln(2)}+1\right)} + b\right)$$

Finally, if we know that $b = 2$, we can compute

```
>   subs(b = 2,");
```

$$4\, n + n\left(-4\left(\frac{1}{2}\right)^{\left(\frac{\ln(n)}{\ln(2)}+1\right)} + 2\right)$$

```
>   simplify(");
```

$$6\, n - 2$$

5.3   Inclusion − Exclusion

We shall begin to look, in this section, at the second of the main two counting techniques covered in Chapter 5 of the text — the principal of inclusion and exclusion. We shall see how to use Maple to solve problems with this technique.

At the heart of the principle of inclusion and exclusion is the formula

$$|A \cup B| = |A| + |B| - |A \cap B|,$$

which says that, for two finite sets A and B, the number of elements in the union $A \cup B$ of the two sets may be found by first adding the sizes $|A|$ of A and $|B|$ of B, and than subtracting the number $|A \cap B|$ of elements common to both A and B, which would otherwise be counted twice. This formula can be generalized to count the number of elements in the union of any finite number of finite sets.

To work with formulae such as these in Maple, it is necessary to learn first how to represent sets in Maple. Since Maple is specially engineered for doing mathematics, this is done quite naturally: to represent a set of elements, we simply list those elements, separated by commas, and enclose the entire construct in braces. For example, to represent the set $\{2, 3, 5\}$ whose members are the numbers 2, 3 and 5, we can use ordinary mathematical notation.

```
>  {2, 3, 5};
```
$$\{2, 3, 5\}$$

In Maple, a set is a first class data structure. You can assign a set to a variable:

```
>  A := {2, 3, 5};
```
$$A := \{2, 3, 5\}$$

Note that Maple's idea of a set corresponds precisely to the mathematical notion. Thus, there is no implied order among the members of a set, nor is there any notion of "multiplicity" for set members. For problems requiring this kind of additional information, other data structures, such as lists or arrays, must be used. We can see this in Maple with the following examples.

```
>  A := {'Alice', 'Bob', 'Eve'};
```
$$A := \{Bob, Eve, Alice\}$$

```
>  B := {'Bob', 'Alice', 'Eve'};
```
$$B := \{Bob, Eve, Alice\}$$

```
>  evalb(A = B);
```
$$true$$

```
>  C := {'Alice', 'Bob', 'Eve', 'Eve'};
```
$$C := \{Bob, Eve, Alice\}$$

```
> evalb(A = C);
```

$$true$$

The `evalb` Maple procedure evaluates a Boolean expression, and returns either `true` or `false`, according to the truth of falsity of the expression. Thus, Maple considers the three sets A, B and C to be the same set. The first example shows that the order in which we list the members of a set is irrelevant, while the second shows that, despite listing the string `'Eve'` twice, Maple only "sees" it once. (Experiment with these examples using lists, which are delimited with brackets rather than braces, to see the difference between sets and lists in Maple.)

To determine the size of a set (the number of objects in it) in Maple, we use the Maple procedure `nops` (think of it as "n operands").

```
> A := { 'Alice', 'Bob', 'Eve'};
```

$$A := \{Bob, Eve, Alice\}$$

```
> nops(A);
```

$$3$$

```
> C := {'Alice', 'Bob', 'Eve', 'Eve'};
```

$$C := \{Bob, Eve, Alice\}$$

```
> nops(C);
```

$$3$$

The set theoretic operators $\cup$ (union) and $\cap$ (intersect) are represented in Maple by writing out their names — union and `intersect`, respectively.

```
> A := {1, 2, 3, 4, 5}: B := {4, 5, 6, 7, 8}:
> A union B;
```

$$\{1, 2, 3, 4, 5, 6, 7, 8\}$$

```
> A intersect B;
```

$$\{4, 5\}$$

In addition, the set theoretic difference is denote by the Maple operator minus.

```
> A minus B;
```

$$\{1, 2, 3\}$$

Let's use the operations to verify the principle of inclusion and exclusion in a particular example.

```
> Flintstones := {'Fred', 'Wilma', 'Pebbles'};
```

$$Flintstones := \{Fred, Wilma, Pebbles\}$$

```
>  Rubbles := {'Barney', 'Betty', 'Bam Bam'};
```
$$Rubbles := \{Barney, Betty, Bam\ Bam\}$$

```
>  Husbands := {'Fred', 'Barney'};
```
$$Husbands := \{Fred, Barney\}$$

```
>  Wives := {'Wilma', 'Betty'};
```
$$Wives := \{Wilma, Betty\}$$

```
>  Kids := {'Pebbles', 'Bam Bam'};
```
$$Kids := \{Pebbles, Bam\ Bam\}$$

If this were a complete census, then the number of children living in Bedrock would be

```
>  nops(Kids);
```
$$2$$

while the number of Bedrock inhabitants who are either Flintstones or children is

```
>  nops(Flintstones union Kids);
```
$$4$$

According to the principle of inclusion and exclusion, this number should also be

```
>  nops(Flintstones) + nops(Kids)
>     - nops(Flintstones intersect Kids);
```
$$4$$

which, of course, it is!

As another example, consider the problem of determining the number of positive integers less than or equal to 100 that are not divisible by either 2 or 11. First, we generate the set of positive integers less than or equal to 100.

```
>  hundred := {seq(i, i = 1..100)}:
```

This shows how you can use Maple's iterator **seq** to generate the members of a set. Next, we get rid of those elements that are divisible by 2:

```
>  A := hundred minus {seq(2 * i, i = 1..100)}:
```

and those that are divisible by 7:

```
>  B := hundred minus {seq(7 * i, i = 1..100)}:
```

(Note the combined use of the **minus** and **seq** operators; they work very conveniently together here.) We are looking for integers that belong to

either or both of A and B, that is, to their union, so we want the size of
the set $A \cup B$, which is

```
> nops(A union B);
```

$$93$$

According to the principle of inclusion and exclusion, this value could also
be computed as

```
> nops(A) + nops(B) - nops(A intersect B);
```

$$93$$

The same principle can be used for larger examples. Here, we outline what
needs to be done to determine the number of positive integers less than
1000 that are indivisible by the primes 2, 3, 5 and 7. To do this, we'll
use the principle of inclusion and exclusion to count those integers less
than 1000 that are divisible by at least one of these four primes, and then
subtract that from 1000.

First, we create the set of positive integers less than or equal to one
thousand.

```
> th := {seq(i, i=1..10^3)}:
```

Now, the integers less than 1000 that are divisible by one of 2, 3, 5 and 7
are those in the union of the sets

```
> th2 := th intersect {seq(2*i, i=1..1000)}:
> th3 := th intersect {seq(3*i, i=1..1000)}:
> th5 := th intersect {seq(5*i, i=1..1000)}:
> th7 := th intersect {seq(7*i, i=1..1000)}:
```

(Note that we do not have to allow the index i to reach 1000 in each of
these, but it is simpler this way, since we will discard the unneeded values
by taking the intersection.) Next, we create the sets of integers that are
divisible by these four primes in pairs.

```
> th_2_3 := th intersect {seq(2*3*i, i=1..1000)}:
> th_2_5 := th intersect {seq(2*5*i, i=1..1000)}:
> th_2_7 := th intersect {seq(2*7*i, i=1..1000)}:
> th_3_5 := th intersect {seq(3*5*i, i=1..1000)}:
> th_3_7 := th intersect {seq(3*7*i, i=1..1000)}:
> th_5_7 := th intersect {seq(5*7*i, i=1..1000)}:
```

We count also those integers less than 1000 that are divisible by the
numbers in triples.

```
> th_2_3_5 := th intersect {seq(2*3*5*i, i=1..1000)}:
> th_2_3_7 := th intersect {seq(2*3*7*i, i=1..1000)}:
> th_2_5_7 := th intersect {seq(2*5*7*i, i=1..1000)}:
> th_3_5_7 := th intersect {seq(3*5*7*i, i=1..1000)}:
```

Finally, we count the numbers less than 1000 that are divisible by all four of 2, 3, 5 and 7.

```
>   th_2_3_5_7 := th intersect {seq(2*3*5*7*i, i=1..1000)}:
```

Now, to compute the number of integers less than 1000 that are divisible by one of 2, 3, 5 and 7, we compute as follows.

```
>   nops(th2) + nops(th3) + nops(th5) + nops(th7);
```
$$1175$$

```
>   " - (nops(th_2_3) + nops(th_2_5) + nops(th_2_7));
```
$$838$$

```
>   " - (nops(th_3_5) + nops(th_3_7) + nops(th_5_7));
```
$$697$$

```
>   " + (nops(th_2_3_5) + nops(th_2_3_7) + nops(th_2_5_7));
```
$$767$$

```
>   " + nops(th_3_5_7) - nops(th_2_3_5_7);
```
$$772$$

Therefore, the number of integers less than 1000 not divisible by 2, 3, 5 or 7 is

```
>   1000 - ";
```
$$228$$

5.4   Generating Functions

Generating functions are a powerful tools for modeling sets of objects and their constructions. For example, if one set of objects is constructed from two others by performing a Cartesian product of two underlying sets, then the generating function for the new set is often just the product of the generating functions for the two underlying sets. Thus, knowing how a set is constructed can help us to construct its generating function.

If you think of the generating functions as polynomials. then every object from the original set is represented in this expansion of the product of the two polynomials by a monomial such as x^5. Several different combinations may lead to an x^5. the coefficient of x^5 in the expanded generating function indicates the number of such objects in the new set.

The coefficients of the expanded generating function form a sequence of numbers - the number of objects in your set of each size. Thus we often refer to a generating function as *the generating function for the sequence* — its coefficients. Appendix 3 of your textbook discusses the use of generating functions. In particular, such sequences can also be described by recurrence relations. Here, we will discuss how to use generating functions to help us to solve those recurrence relations.

The **generating function** $g(x)$ for a sequence $\{r_n\}$ is the formal power series

$$\sum_{k=0}^{\infty} r_k x^k = r_0 + r_1 x + r_2 x^2 + r_3 x^3 + \cdots + r_n x^n + \cdots$$

It is called *formal* because we are not at all interested in evaluating it as a function of x. Our entire focus is on finding formulae for its coefficients. In particular, this means that there are no convergence issues to be investigated.

Maple provides extensive facilities for manipulating formal power series (that is, generating functions). They belong to the Maple package **powseries**, so to access these facilities, you must tell Maple to load this package.

```
>  with(powseries);
```

    [*compose, evalpow, inverse, multconst, multiply, negative, powadd,*
       powcos, powcreate, powdiff, powexp, powint, powlog, powpoly, powsin,
       powsolve, powsqrt, quotient, reversion, subtract, tpsform]

The first thing we need to do is learn how to create a power series. For this, Maple provides the function **powcreate**. It takes as arguments a sequence of equations defining the general coefficient. The equations specify a way of computing the kth coefficient in $\sum_{k=0}^{\infty} a_k x^k$. For example, the formal exponential function, which has the power series representation

$$\exp(s) = \sum_{n=0}^{\infty} \frac{s^n}{n!},$$

can be created in Maple by issuing the call

```
>  powcreate(e(n) = 1/n!);
```

What makes this especially useful for working with recurrence relations is that the general coefficient need not be specified in closed form (as it was above). You can specify a recurrence relation satisfied by the coefficients, together with sufficiently many initial conditions to guarantee a unique solution to the recurrence.

Let's see an example of this. To create the generating function for the Fibonacci sequence, which is defined by the recurrence relation

$$f_n = f_{n-1} + f_{n-2} \quad \text{and} \quad f(0) = 1, f(1) = 1,$$

we can enter

```
>   powcreate(f(n) = f(n - 1) + f(n - 2), f(0) = 1, f(1) = 1);
```

Now, the only interesting information in a generating function is the sequence of its coefficients. Maple provides a way to access an arbitrary coefficient in a formal power series. This is done as follows. To Maple, each formal power series is, in fact, a procedure, which takes integer arguments. The value returned by a formal power series when given an integer n as argument is the coefficient of x^n. So, for example, the fifth Fibonacci number can be produced by calling the formal power series f above with '5' as argument.

```
>   f(5);
```

$$8$$

In fact, the general coefficient may be obtained by passing the special argument _k

```
>   f(_k);
```

$$f(\_k - 1) + f(\_k - 2)$$

To display a generating function, it is best to use the Maple function tpsform. This procedure converts a formal power series into a truncated power series of the specified degree. For instance, to display the first ten terms of the generating function for our Fibonacci sequence, we can use tpsform, as follows.

```
>   tpsform(f, x, 9);
```

$$1 + x + 2\,x^2 + 3\,x^3 + 5\,x^4 + 8\,x^5 + 13\,x^6 + 21\,x^7 + 34\,x^8 + O(x^9)$$

Generating functions are more than just a convenient way to represent numerical sequences and their associated sets of objects. They are a powerful tool for solving recurrence relations, as well as other kinds of counting problems. This power stems from our ability to manipulate them, more or less, like ordinary power series from Calculus and to interpret those manipulations in terms of their action on the sets.

Just as is done in Calculus with ordinary power series, generating functions may be added, multiplied, multiplied by scalars and polynomials, composed, evaluated, and even differentiated and integrated. It is important to recognize that we are speaking here of *formal* differentiation and integration — there are no limits to worry about.

It is even more important to associate these "algebraic" operations with combinatorial operations that you might carry out on the set of objects implicitly represented by the generating function. For example, taking the union of two disjoint sets of objects corresponds to adding their generating functions. Each of the operations are often best thought of in terms of their affect on the "monomials" that represent the individual objects of the underlying set of objects. For example, If a single object made of of five sub-objects is represented by x^5 then there is exactly 5 ways of choosing one of those sub-objects for removal. The set of objects produced by doing this in all possible ways would be represented by $5x^4$. Thus, in a very real sense, this combinatorial operation of breaking up a single object in this way corresponds to the familiar operation of differentiation on its generating function.

All the most common operations you can carry out on ordinary power series have useful combinatorial interpretations and can be carried out on our formal power series. In each case, we can specify what such an effect will have on the coefficient of the series. Maple provides facilities for performing all of these manipulations, and more.

These facilities are best demonstrated by working through an example. We will use Maple to solve the Fibonacci recurrence with generating functions.

If we multiply both sides of the Fibonacci recurrence

$$f_n = f_{n-1} + f_{n-2}$$

by x^n, then we obtain

$$f_n x^n = f_{n-1} x^n + f_{n-2} x^n.$$

Now summing from $n = 1$ yields

$$\sum_{n=1}^{\infty} f_n x^n = \sum_{n=1}^{\infty} f_{n-1} x^n + \sum_{n=1}^{\infty} f_{n-2} x^n.$$

The left side of this equation differs from the generating function by only the first term (in which $n = 0$), and the sums on the right side can be factored so we obtain

$$g(x) - 1 = xg(x) + x^2 g(x).$$

Now, solving this equation for $g(x)$ produces

$$g(x) = \frac{-1}{x^2 + x - 1}.$$

5.5   Computations and Explorations

This section will present some Maple solutions to a few of the problems mentioned in the *Computer Projects* and *Computations and Explorations* section of your textbook. We shall not always present here a complete solution; in a few cases, we just suggest one or two things for you to try, and leave the detailed implementation up to you.

2 The next problem we shall consider is that of determining the smallest Fibonacci number that exceeds one million, one billion and one trillion.

Solution: We can solve this quite easily within Maple, using a simple while loop. First, however, we must make certain that we get the correct fibonacci function.

```
>  with(combinat):
```

This defines the correct version of the Maple function fibonacci for us. There is another function, also called fibonacci in the linalg package, but it is the wrong function.

The idea here is to loop over the index to the Fibonacci sequence until the value of the sequence reaches a specified limit (say, one million). The while loop construct in Maple is ideally suited to this sort of application.

```
>  count := 1;  # initialize a counter
                    count := 1
```

```
>  while fibonacci(count) <= 1000000 do
>    count := count + 1;
>  od:
>  print(fibonacci(count));
                   1346269
```

We can see which Fibonacci number gives us this value by checking the value of the variable count.

```
>  count;
```

It is probably also a good idea to check our logic, and see that the previous Fibonacci number really is less than 1000000.

```
>  fibonacci(count - 1);
```
$$832040$$

Now, we are supposed to check this for a few more values even larger than one million. However, once you have tried two or three, you will certainly want to try more, so it is probably a good idea to wrap this little `while` loop up inside of a function (which we'll call `BigFib` here).

```
>  BigFib := proc(n)
>    # compute smallest Fibonacci number exceeding n
>    local k;
>
>    with(combinat);
>    k := 1;
>
>    while fibonacci(k) <= n do
>        k := k + 1;
>    od;
>    print(fibonacci(k));
>  end:
```

To make our function work correctly, we have called `with(combinat)` in the body of the function to ensure that we get the correct version of the `fibonacci` function. (This could also be achieved by using the long calling syntax `combinat[fibonacci]` for the function.) Now it is quite simple to compute the smallest Fibonacci number exceeding a given number.

```
>  BigFib(1000000000);
```
$$1134903170$$

```
>  BigFib(1000000000000);
```
$$1548008755920$$

```
>  BigFib(10^10);
```
$$12586269025$$

3 Find as many prime Fibonacci numbers as you can.

Solution: Using Maple, this sort of problem becomes very straightforward; We can simply use the Maple procedure `fibonacci`, from the `combinat` package to generate Fibonacci numbers, and we can use the `isprime` function to test each for primality. Despite being very simple, we'll wrap this up in a procedure, so that we can call it with different arguments that determine how many Fibonacci numbers will be tested.

```
> PrimeFib := proc(n)
>   local i,          # loop index
>         t,          # temporary variable
>         prime_fib; # list of prime Fibonacci numbers; returned
>   prime_fib := NULL;
>   for i from 1 to n do
>       t := combinat[fibonacci](i);
>       if isprime(t) then prime_fib := prime_fib, t; fi;
>   od;
>   RETURN(prime_fib);
> end:
```

Here, to save space, we test only the first 100 Fibonacci numbers.

```
> PrimeFib(100);
```
$$2, 3, 5, 13, 89, 233, 1597, 28657, 514229, 433494437, 2971215073,$$
$$99194853094755497$$

Note that, since we use isprime, our list is not certain to consist solely of prime numbers, as isprime uses a probabilistic primality test.

Another approach that you may consider trying is to construct two lists: one containing the list of Fibonacci numbers up to some point, and the other containing the sequence of primes, generated using the ithprime function (which is not probabilistic). Then traverse the two lists to extract any members they have in common. This approach has the advantage that it avoids the use of the probabilistic primality test used by isprime.

7 Find all the prime numbers not exceeding 1000, using the sieve of Eratosthenes.

Solution: Implementing Eratosthenes' sieve is a nontrivial exercise in any programming language, but Maple makes it easier than most. The implementation that we provide here is designed to follow the description given in the textbook fairly closely.

The sieve produces a list of all the prime numbers not exceeding a given positive integer n. We shall model the list of integers from 1 to n by a boolean valued array isprime. The ith entry isprime[i] will have the value **true** if i is a prime number, and false otherwise. At the beginning of the algorithm, all entries are initialized to **false**. This corresponds to having written down the list of numbers from 1 to n, but not having yet crossed any out. To "cross out a number", we set its value in the array isprime to false. Progressing through the algorithm detects non-primality, and entries will be marked **false** as they are discovered to be composite.

Our program consists principally of three `for` loops. The first simply initializes the array `isprime`, while the third `for` loop prints out the results. The sieve itself is the middle `for` loop, as described in the textbook.

We use three new functions in the code. The `array` function simply creates an uninitialized array. (Use Maple's help facility to learn more about `array`.) The function `isqrt` produces an integer approximation of the square root of its argument. The most interesting new feature is the call to `type`, which tests whether or not its first argument has the type of its second argument. Here, it is being used to test whether the result of a division is an integer, which effectively determines whether one integer divides another. Another way to accomplish this would be to use the `irem` function, it produces the remainder after dividing its first argument by its second argument.

```
>  irem(5,2);
```

$$1$$

```
>  irem(6,2);
```

$$0$$

The line which reads if `type(j/i, integer)` then, in our code, could be replaced with if `irem(j,i) = 0` then.

Here is the code.

```
>  Esieve := proc(n)
>     local i,j,      # loop indices
>         isPrime,    # array of booleans
>         prime_list, # list of primes
>         sqrtn;      # integer approx. of sqrt(n)
>     #options trace;
>     # initialize the array
>     isPrime := table();
>     isPrime[1] := false;
>     for i from 2 to n do
>         isPrime[i] := true
>     od;
>     # get an integer approximation to
>     # the square root of the
>     # argument 'n' (add 1 for safety).
>     sqrtn := 1 + isqrt(n);
>     # the actual sieve
>     for i from 1 to sqrtn do
>         # skip it if it is not prime
>         if isPrime[i] then
>             for j from i+1 to n do
>                 # test whether i divides j
>                 #if type(j/i, integer) then
>                 if irem(j,i) = 0 then
>                     isPrime[j] := false
```

```
>                      fi;
>                 od;
>           fi;
>      od;
>      # convert the list of booleans to a list of primes
>      prime_list := NULL;
>      for i from 1 to n do
>                  if isPrime[i] then
>                          prime_list := prime_list, i;
>                  fi;
>      od;
>      RETURN(prime_list);
> end:
```

Now try it out!

```
> Esieve(10);
```

$$2, 3, 5, 7$$

```
> Esieve(100);
```

$2, 3, 5, 7, 11, 13, 17, 19, 23, 29, 31, 37, 41, 43, 47, 53, 59, 61, 67, 71, 73,$
$79, 83, 89, 97$

```
> Esieve(1000);
```

$2, 3, 5, 7, 11, 13, 17, 19, 23, 29, 31, 37, 41, 43, 47, 53, 59, 61, 67, 71, 73,$
$79, 83, 89, 97, 101, 103, 107, 109, 113, 127, 131, 137, 139, 149, 151,$
$157, 163, 167, 173, 179, 181, 191, 193, 197, 199, 211, 223, 227, 229,$
$233, 239, 241, 251, 257, 263, 269, 271, 277, 281, 283, 293, 307, 311,$
$313, 317, 331, 337, 347, 349, 353, 359, 367, 373, 379, 383, 389, 397,$
$401, 409, 419, 421, 431, 433, 439, 443, 449, 457, 461, 463, 467, 479,$
$487, 491, 499, 503, 509, 521, 523, 541, 547, 557, 563, 569, 571, 577,$
$587, 593, 599, 601, 607, 613, 617, 619, 631, 641, 643, 647, 653, 659,$
$661, 673, 677, 683, 691, 701, 709, 719, 727, 733, 739, 743, 751, 757,$
$761, 769, 773, 787, 797, 809, 811, 821, 823, 827, 829, 839, 853, 857,$
$859, 863, 877, 881, 883, 887, 907, 911, 919, 929, 937, 941, 947, 953,$
$967, 971, 977, 983, 991, 997$

The last computation answers the problem from the text.

8 Compute the number of onto functions from one finite set to another, given their sizes.

Solution: First, convince yourself that the sizes of the domain and codomain are the only parameters required to compute this value. We have a very convenient formula, derived using the principle of inclusion and exclusion for this number, given by

$$\sum_{k=0}^{n-1}(-1)^{k}C(n,k)(n-k)^{m}$$

which is the number of onto functions from a set of m elements to a set of n elements. This formula is derived in the text (see Page 344). The only input required in this formula are the integer parameters m and n which represent the sizes, respectively, of the domain and the codomain. The index k of summation can be treated as a local variable in a Maple procedure, since it is used only temporarily during the calculation. So, our Maple function for computing this formula, can take the integers m and n as arguments, and will return the number of onto functions from a set with m elements to a set with n elements. Here is the function.

```
>   with(combinat):
>   OntoFunctions := proc(m, n)
>     local k, s;
>
>     if m < n then
>         RETURN(0);
>     fi;
>     s := sum(
>       (-1)^k * binomial(n,k) * (n - k)^m,
>       k = 0..(n-1)
>     );
>     RETURN(s);
>   end:
```

The if statement is necessary — and makes perfect sense, mathematically — because there are no onto functions from a set to another set which is larger; in other words, the number of onto functions from a smaller set to a larger set is 0, which is precisely what our program returns.

```
>   OntoFunctions(4,9);
```
$$0$$

The local variable s is not necessary at all; its only purpose here is to improve the readability of the program.

```
>   OntoFunctions(5,4);
```
$$240$$

```
>   OntoFunctions(100,20);
```
11238195910319657928539447038143170285517894975095769\
49629431900741309191395982833493646419629819250889018\
231616326106793426944000

It is probably obvious that the number of onto functions from one set to another increases with the sizes of either the domain or the range. Experiment with this function to see if you can determine whether an increase in the size of the domain or the range makes the greatest impact on the number of onto functions.

10 The final problem we will look at here is problem 10 from the Computations and Explorations section of the text. This problem asks you to compute the probability that a permutation of n objects is a derangement, for all positive integers n such that $n \le 20$, and determine how quickly these probabilities approach the number e.

Solution: To solve this problem, we need to know the formula which gives the number of derangements of n objects, namely,

$$D_n = n! \left[1 - \frac{1}{1!} + \frac{1}{2!} - \frac{1}{3!} + \cdots + (-1)^n \frac{1}{n!} \right].$$

The total number of permutations of n objects if, of course, $n!$, so the probability that one of them is a derangement is just the ratio $D_n/n!$, which is given by the expression

$$1 - \frac{1}{1!} + \frac{1}{2!} - \frac{1}{3!} + \cdots + (-1)^n \frac{1}{n!}.$$

A very simple Maple function will compute these values for us.

```
>  DerProb := proc(n::integer)
>     local k;
>     RETURN(sum((-1)^k * (1/k!), k = 0 .. n));
>  end:
```

To test it out, we try to answer the original question, except that, to save space, we only test values less than or equal to 5.

```
>  seq(DerProb(i),i=1..5);
```

$$0, \frac{1}{2}, \frac{1}{3}, \frac{3}{8}, \frac{11}{30}$$

How do these numbers differ from e^{-1}?

```
>  for i from 1 to 5 do
>     print(evalf(E * DerProb(i) - 1));
>  od;
```

$$-1.$$
$$.5000000000 \, E - 1.$$
$$.3333333333 \, E - 1.$$
$$.3750000000 \, E - 1.$$
$$.3666666667 \, E - 1.$$

You can try this for values up to 20, or even greater.

5.6   Exercises/Projects

Exercise 1. Use Maple to solve the following recurrence relations.

(a) $r_n = r_{n-1} - r_{n-2}$, $r_1 = 1$, $r_2 = 1$;

(b) $r_n = 15r_{n-1} + r_{n-2}/2$, $r_1 = \frac{23}{22}$, $r_2 = \frac{7}{2}$.

Exercise 2. Use Maple to solve each of the recurrence relations in Exercise 5 on Page 315 of the text.

Exercise 3. Write a general solver in Maple for linear, homogeneous recurrence relations with constant coefficients of degree three. Assume that the roots of the characteristic polynomial of the recurrence relation are distinct. Better yet, have your procedure check that this is, in fact, the case.

Exercise 4. Use Maple to investigate the behavior of the limit

$$\lim_{n \to \infty} = \frac{\varphi_n}{\psi_n}$$

where φ_n is defined to be the number of *prime* Fibonacci numbers less than or equal to n, and ψ_n is defined to be the number of Fermat numbers less than or equal to n.

Exercise 5. Use Maple to find the number of square free integers less than 100000000.

Exercise 6. Use Maple to find the number of onto functions from a set with 1000000 elements to a set with 1000 elements.

Exercise 7. To generate the *lucky numbers* start with the positive integers and delete every second integer in the list starting with the integer 1. Other than 1 the smallest integer left is 3; continue by deleting every third integer left, starting the count with 1. The next integer left is 7; continue by deleting every seventh integer left, starting with 1. Continue the process where at each stage every kth integer is deleted where k is the smallest integer left other than one. The integers that remain are the lucky numbers. Develop a Maple procedure generating the lucky numbers.

Exercise 8. Can you make any conjectures about lucky numbers by looking at a list of the first 1000 of them? For example, what sort of conjectures can you make about twin lucky numbers? What evidence do you have for your conjectures?

6 Relations

In this chapter we will learn how to use Maple to work with binary and n-ary relations. We explain how to use Maple to represent binary relations using sets of ordered pairs, using representations using zero-one matrices, and using directed graphs. We show how to use Maple to determine whether a relation has various properties using these different representations.

We describe how to compute closuresof relations. In particular, we show how to find the transitive closureof a relation using two different algorithms, and we compare the time required to use these algorithms. After explaining how to use Maple to work with equivalence relations,we show how to use Maple to work with partial orderings. We show how to use Maple to do topological sortingand to determine whether a partial ordering is a lattice. We conclude by showing how to use Maple to find covering relationsof partial orderings.

6.1 An Introduction to Relations in Maple

The first step in understanding relations and their manipulation in Maple is to determine how to represent relations in Maple. The reader will note that there is no specific `relations` package present in Maple, and hence, the implementation and representation of relations in Maple can take the most convenient form for the question at hand. Possible representations of relations in Maple include sets of ordered pairs, zero-one matrices or directed graphs, among many others. For this chapter, we will examine ordered pair representations and zero-one matrices, as well as the directed graph representation.

First, we shall represent relations as ordered pairs. To this end, we construct a structured `typerel` for relations. (Note that we cannot use the name `relation`, since that type is already used by the Maple library.) According to the definition, a relationis just a set of ordered pairs of any type of object whatsoever. The following Maple type reflects this definition.

```
>   'type/pair' := [anything, anything];
```
$$type/pair := [anything, anything]$$

```
>   'type/rel' := set(pair);
```
$$type/rel := set(pair)$$

This is useful, since it allows us to ensure that, when we pass arguments to functions, we have used the correct data type for input. Since our type `rel` is *structured*, it is much easier to write `rel` than to write `set([anything, anything])` after each argument that is to be interpreted as a relation. This can also be accomplished by doing such checking "by hand" inside each function, before the real processing begins, but Maple provides this automatic type checking for us and results in faster and, more importantly for us, more readable code.

For a specific example, suppose we wish to establish a relation that is defined in terms of numerical constraints, as in Example 4 on page 357 of the text. In the text example, we need to create a relation R on the domain $A = \{1, 2, 3, 4\}$, such that $R = \{(a, b) | a \text{ divides } b\}$. We will construct this relation by examining every possible ordered pair of elements, and admitting ordered pairs into R if, and only if, the ordered pair satisfies the appropriate condition. We shall call R `DividesRelation`.

```
> DividesRelation := proc(A::set(integer))
>    local i, j, temp, R;
>    R := {};
>    for i in A do
>      for j in A do
>        if (gcd(i,j) = i) then
>          R := R union {[i, j]};
>        fi;
>      od;
>    od;
>    RETURN(R);
> end:
> DividesRelation({1,2,3,4});
```
$$\{[1, 1], [1, 2], [1, 3], [1, 4], [2, 2], [2, 4], [3, 3], [4, 4]\}$$

For convenience, we also define the following variation.

```
> DivRel := proc(n::posint)
>    local i;
>    DividesRelation({seq(i,i=1..n)});
> end:
```

This procedure constructs the "divides" relation on the set of all integers in the set $\{1, 2, 3, \ldots, n\}$.

It will be convenient to have at hand the following procedure for creating the "dual" or "opposite" relation of a given relation.

```
> DualRelation := proc(R::rel)
>    local u;
>    map(u -> [u[2], u[1]], R);
> end:
```

This simply reverses all the pairs that belong to the relation.

6.2 Determining Properties of Relations using Maple

Maple can be used to determine if a relation has a particular property, such as reflexivity, symmetry, antisymmetry or transitivity. This can be accomplished by creating Maple procedures that take as input the given relation, examining the elements of the relation, and determining whether the relation satisfies the given property.

Since we shall use it repeatedly, it will be convenient to have a routine that will extract for us the domainof any relation. We simply collect together all the points that occur as either a first or second entry in some pair in the relation. (Not that, strictly speaking, this need not equal the domain of the relation, since there may, in fact, exists points in the domain that are not R-related to any other point in the domain. It might be better to call this the "effective domain" of the relation.)

```
>   DomainRelation := proc(R::rel)
>     RETURN(map(u->op(1,u), R)
>            union map(u->op(2,u), R));
>   end:
```

First, we examine how to determine if a relation is reflexive.

```
>   IsReflexive := proc(R::rel)
>     local is_reflexive,    # return value
>           u;               # index into Dom(R)
>     is_reflexive := true;
>     for u in DomainRelation(R) do
>        is_reflexive := is_reflexive and member([u,u], R);
>     od;
>     RETURN(is_reflexive);
>   end:
```

```
>   R1 := {[1,1], [1,2], [2,1], [2,2], [3,4], [4,1], [4,4]}:
>   R2 := {[1,1], [1,2], [2,1]}:
>   IsReflexive(R1);
```

$$false$$

```
>   IsReflexive(R2);
```

$$false$$

We shall examine the symmetricand antisymmetricproperties in the next two procedures. To determine whether a relation is symmetric we shall simply use the definition; that is, we check whether, for each member (a, b) in a relation, the pair (b, a) is also a member of the relation. If we discover a pair (a, b) in the relation for which the pair (b, a) is not in the relation, then we know that the relation is not symmetric. Otherwise, it must be symmetric. This is the logic employed by the following procedure.

```
>  IsSymmetric := proc(R::rel)
>     local i, is_symmetric;
>     is_symmetric := true;
>     for i from 1 to nops(R) do
>        if not member( [ R[i][2], R[i][1] ], R) then
>           is_symmetric := false;
>        fi;
>     od;
>     RETURN(is_symmetric);
>  end:
```

To determine whether a given relation R is antisymmetric, we again use the definition. Remember that for a relation R to be antisymmetric, it must have the property that, whenever a pair (a, b) belongs to R, and the pair (b, a) also belongs to R, then we must have that $a = b$. To check this, we simply loop over all the members $u = (a, b)$ of R, and see whether the opposite pair (b, a) belongs to R. If it does, and if $a \neq b$, then R cannot be antisymmetric; otherwise, it is.

```
>  IsAntiSymmetric := proc(R::rel)
>     local u; # index into R
>
>     for u in R do
>        if member([op(2,u), op(1,u)], R)
>           and op(1, u) <> op(2, u) then
>              RETURN(false);
>        fi;
>     od;
>     RETURN(true);
>  end:
```

We now use our procedures to determine which of the relations defined earlier are symmetric or antisymmetric.

```
>  IsSymmetric(R1); IsSymmetric(R2);
```
$$false$$
$$true$$

```
>  IsAntiSymmetric(R1); IsAntiSymmetric(R2);
```
$$false$$
$$false$$

```
>  R3 := {[1,1], [1,2], [1,4], [2,1],
>         [2,2], [3,3], [4,1], [4,4]}:
>  R4 := {[2,1], [3,1], [3,2], [4,1],
>         [4,2], [4,3]}:
>  IsAntiSymmetric(R3); IsAntiSymmetric(R4);
```
$$false$$
$$true$$

To decide whether a relation R is transitive, we must check whether (a, c) belongs to R if there exist pairs (a, b) and (b, c) that belong to R. We do this by examining all pairs (a, b) in R, together with all elements x of the domain D of R, to see whether a pair (b, x) exists in R for which the pair (x, b) is not in R. If we find such a combination, then the relation R cannot be transitive; otherwise, R is transitive.

```
> IsTransitive := proc(R::rel)
>    local DomR,   # domain of R
>          u,v,w;# indices into DomR
>    for u in DomR do
>      for v in DomR do
>        for w in DomR do
>          if (member([u,v], R) and member([v,w], R)
>                        and not member([u,w], R)) then
>              RETURN(false);
>          fi;
>        od;
>      od;
>    od;
>    RETURN(true);
> end:
> IsTransitive(R1); IsTransitive(R2);
                          true
                          true

> IsTransitive(R3); IsTransitive(R4);
                          true
                          true
```

6.3 n-ary Relations in Maple

Using Maple, we can construct an n-ary relation where n is a positive integer. The format for the n-ary relation expression in Maple is similar to that of the 2-ary relation in Maple. For example, consider the following 4-ary relation that represents student records.

```
> M1:={[Adams, 9012345,'Politics', 2.98],
>      [Woo, 9100055, 'Film Studies', 4.99],
>      [Warshall, 9354321, 'Mathematics', 3.66]
> }:
```

The first field represents the name of the student, the second field is the student ID number, the third field is the students home department and, finally, the last record stores the students grade point average.

As an example of how to use n-ary relations, we shall construct a general procedure for computing a given projection of a relation. The procedure takes as input a set of *n*-tuples as the relation along with the image that is to be projected upon. The output of this procedure is a set of *m*-tuples.

```
>  MakeProjection := proc(R::set, P::list)
>     local i, j, S, temp_list;
>     S := {};
>     for i from 1 to nops(R) do
>       temp_list := [];
>       for j from 1 to nops(P) do
>         temp_list := [op(temp_list), R[i][P[j]]];
>       od;
>       S := S union {temp_list};
>     od;
>     S;
>  end:
```

We now examine this projection procedure on the 4-ary relation that was constructed earlier in this section.

```
>  MakeProjection(M1, [3,4,1]);
```
$$\{[Politics, 2.98, Adams], [Film\ Studies, 4.99, Woo],$$
$$[Mathematics, 3.66, Warshall]\}$$

```
>  MakeProjection(M1, [2,4]);
```
$$\{[9012345, 2.98], [9100055, 4.99], [9354321, 3.66]\}$$

We now move from constructing projections to the construction of joins of relations. The join operation has applications in database commands when tables of information need to be combined in a meaningful manner. The join operation that we will implement in Maple follows this pseudocode outline:

1. Input two relations A and B, and an nonnegative integer parameter p

2. Examine each element of A, and determine the last p fields of each element x

3. Examine all elements, y, of relation B to determine if the first p fields of y match the last p fields of element x

4. Upon finding a match of an element in A and an element in B, we combine these elements, placing the result in C, which is returned as output.

```
>  MakeJoin := proc(p, A, B)
>     local i, j, k, C, list_A, list_B, x,
>     ret_elem, is_done;
>     list_A := [];
>     list_B := [];
>     C := {};
>     for i from 1 to p do
>       list_B := [op(list_B), i];
```

```
>        list_A := [nops(B[1])-i, op(list_A)];
>      od;
>      for i from 1 to nops(A) do;
>        is_done := false;
>        x := MakeProjection({A[i]}, list_A);
>        j := 1;
>        while j <= nops(B) and is_done = false do
>          if MakeProjection({B[j]}, list_B) = x then
>          ret_elem := A[i];
>          for k from p+1 to nops(B[j]) do
>            ret_elem := [op(ret_elem), B[j][k]];
>          od;
>          is_done := true;
>          fi;
>          j := j+1;
>        od;
>      C := C union {ret_elem};
>      od;
>      C;
>    end:
```

We examine how this procedure works on the example defined in the textbook on Pages 371–372, involving courses that Professors are teaching in determining where they are located on campus.

```
>   A:={[Cruz, Zoology, 335],
>       [Cruz, Zoology, 412],
>       [Farber, Psychology, 501],
>       [Farber, Psychology, 617],
>       [Grammer, Physics, 544],
>       [Grammer, Physics, 551],
>       [Rosen, Computer, 518],
>       [Rosen, Mathematics, 575]
>   }:
>   B:={[Computer, 518, N521, 14],
>       [Mathematics, 575, N502, 15],
>       [Mathematics, 611, N521, 16],
>       [Physics, 544, B505, 16],
>       [Psychology, 501, A100, 15],
>       [Psychology, 617, A110, 11],
>       [Zoology, 335, A100, 9],
>       [Zoology, 412, A100, 8]
>   }:
>   MakeJoin(2, A, B);
```

$\{[Cruz, Zoology, 335, A100, 9], [Cruz, Zoology, 412, A100, 8],$

$\quad [Farber, Psychology, 501, A100, 15],$

$\quad [Farber, Psychology, 617, A110, 11],$

$\quad [Grammer, Physics, 544, B505, 16], [Rosen, Computer, 518, N521, 14],$

$\quad [Rosen, Mathematics, 575, N502, 15]\}$

6.4 Representing Relations as Digraphs and Zero-One Matrices

As was stated earlier in this chapter, Maple allows us to represent and manipulate relations in a variety of ways. We have seen how the `combinat` package of Maple allows the Cartesian product to be used to generate relations and, in this section, we will use the `networks` and `linalg` packages to represent and manipulate relations. We have explained how to use Maple to work with relations using the representation of relations as sets of ordered pairs. In this section we will show how to use Maple to work with relations using two alternate methods to represent relations. First, we shall explain how to represent relations as directed graphs; this requires the use of the Maple `networks` package. Second, we will show how to represent relations using zero-one matrices; this requires the use of the Maple linalg package. We will see that using these alternate representations of relations allows us to use a wider range of Maple's capabilities to solve problems involving relations.

Representing Relations Using Directed Graphs

We begin by examining how to represent relations using directed graphs in Maple, with the help of the `networks` package. To begin, we will load the `networks` package;

```
>  with(networks):
```

Now, we can convert our relations that were represented in ordered pair format into a directed graph using the following simple algorithm, called MakeDigraph.

```
>  MakeDigraph := proc(A::set,R::set)
>    local G;
>    new(G);
>    addvertex(A, G);

>    addedge(R, G);

>    RETURN(G);
>  end:
>  G1 := MakeDigraph({1,2,3,4},R1);
```
$$G1 := G$$

```
>  R1;
```
$$\{[1, 1], [1, 2], [2, 2], [4, 4], [2, 1], [3, 4], [4, 1]\}$$

The procedure **new** from the **networks** package creates an instance of a graph, and **addedge** does exactly what its name suggests: it adds an edge to the graph that is its second argument. (A fuller discussion of these routines will be presented in Chapter 7.)

We can now use this graphical representation of the relation R1 to deduce whether or not it is transitive. To do this, we use the *all-pairs shortest path* algorithm, denoted as **allpairs** in Maple. Specifically, we wish to determine that if an edge (a, b) and an edge (b, c) occur in the given relational graph, then edge (a, c) must occur. Hence, we use the following pseudocode outline.

1. Input a graph G, which represents a relation R.

2. Execute the all pairs shortest path algorithm on G, which returns the shortest path between any two points.

3. If there is a pair of elements that has finite length that is greater than 1, then we know that the graph is not reflexive.

4. Otherwise, we have all (finite) lengths between elements one, so that if (a, b) is a pair and (b, c) is a pair, then the distance from a to c is 1 (since this is the only possible length, since we eliminated finite lengths that are greater than 1 in step 3), so (a, c) must be an edge, and hence (a, c) is in the relation R.

5. Output the value of the decision from steps 3 and 4.

The implementation of this pseudocode is the following;

```
>   IsTransitive_G := proc(G::graph)
>     local i, j, S, T, is_trans;
>     is_trans := true;
>     T := allpairs(G);
>     S := vertices(G);
>     for i from 1 to nops(S) do;
>       for j from 1 to nops(S) do;
>         if T[S[i],S[j]]>1 and T[S[i],S[j]]<infinity then
>           is_trans := false
>         fi;
>       od;
>     od;
>     is_trans;
>   end:
>   IsTransitive_G(G1);
```

$$false$$

```
>   R2;
```

$$\{[1, 1], [1, 2], [2, 1]\}$$

```
> IsTransitive_G(MakeDigraph({1,2,3,4},R2));
```
true

You should examine other ways to manipulate the graphical representation of relations in Maple. In particular, once you have studied Chapter 7, you can explore a variety of ways to manipulate graphs, and see how information about relations can be extracted from them.

Representing Relations Using Zero-One Matrices

We now will consider the representation of relations by zero-one matrices. To begin, since we will be working with matrices, we need to indicate that we shall be using functions from the Maple `linalg` package. Specifically, we will need to use the `matrix` function, and related matrix operations, of the `linalg` package.

```
> with(linalg):
```

We now provide a Maple procedure that finds the zero-one matrix representation of a relation, given the ordered pairs in this relation. The pseudocode for this algorithm is as follows;

1. Input the set R and domain D.
2. For each pair (i, j) in D, we determine whether (i, j) is in R.
3. If the pair is in R, we place a 1 entry at the position representing (i, j) in matrix M. Otherwise, we place a 0 at the position representing (i,j) in matrix M.
4. Return M.

Translating this algorithm into Maple code results in the following procedure.

```
> MakeMatrix := proc (R::set, D::set)
>    local i, j, L;
>    L := [];
>    for i from 1 to nops(D) do
>      for j from 1 to nops(D) do
>        if member([i,j],R) then
>           L := [op(L), 1] else L := [op(L),0];
>        fi;
>      od;
>    od;
>    evalm(matrix(nops(D), nops(D), L));
> end:
```

Next, we will convert the relations defined earlier in this chapter from their set form to the representative zero-one form.

```
> m1:=MakeMatrix(R1,{1,2,3,4});
```

$$m1 := \begin{bmatrix} 1 & 1 & 0 & 0 \\ 1 & 1 & 0 & 0 \\ 0 & 0 & 0 & 1 \\ 1 & 0 & 0 & 1 \end{bmatrix}$$

```
> m2:=MakeMatrix(R2,{1,2,3,4});
```

$$m2 := \begin{bmatrix} 1 & 1 & 0 & 0 \\ 1 & 0 & 0 & 0 \\ 0 & 0 & 0 & 0 \\ 0 & 0 & 0 & 0 \end{bmatrix}$$

```
> m3:=MakeMatrix(R3, {1,2,3,4});
```

$$m3 := \begin{bmatrix} 1 & 1 & 0 & 1 \\ 1 & 1 & 0 & 0 \\ 0 & 0 & 1 & 0 \\ 1 & 0 & 0 & 1 \end{bmatrix}$$

```
> m4:=MakeMatrix(R4,{1,2,3,4});
```

$$m4 := \begin{bmatrix} 0 & 0 & 0 & 0 \\ 1 & 0 & 0 & 0 \\ 1 & 1 & 0 & 0 \\ 1 & 1 & 1 & 0 \end{bmatrix}$$

Now that we have the zero-one matrix representations of these relations, we can use these matrices to determine whether the relations are reflexive, symmetric and antisymmetric. In this form, it is somewhat easier to determine whether a given relation has one of these properties. Here, for example, is a Maple procedure that determines whether a relation is reflexive, using its zero-one matrix representation.

```
> IsReflexive_M:= proc(M::matrix)
>     local i, is_reflex;
>     is_reflex := true;
>     for i from 1 to coldim(M) do
>         if M[i,i] = 0 then
>             is_reflex := false;
>         fi;
>     od;
>     is_reflex;
```

```
> end:
> IsReflexive_M(m1);
```
                              false

```
> IsReflexive_M(m3);
```
                              true

Here also are matrix versions of IsSymmetric and IsAntiSymmetric.

```
> IsSymmetric_M := proc(M::matrix)
>   local i,j; # row and column indices
>   if rowdim(M) <> coldim(M) then
>     # must be square
>     RETURN(false);
>   fi;
>   for i from 1 to rowdim(M) do
>     for j from 1 to i-1 do
>       if M[i,j] <> M[j,i] then
>         RETURN(false);
>       fi;
>     od;
>   od;
>   RETURN(true);
> end:
```

The code for checking antisymmetry is similar.

```
> IsAntiSymmetric_M := proc(M::matrix)
>   local i,j;
>   if rowdim(M) <> coldim(M) then
>     RETURN(false);
>   fi;
>   for i from 1 to rowdim(M) do
>     for j from 1 to i - 1 do
>       if M[i,j] = 1 and M[i,j] = M[j,i] then
>         RETURN(false);
>       fi;
>     od;
>   od;
>   RETURN(true);
> end:
```

Again, we determine if the relations represented in zero-one matrix form
are symmetric or antisymmetric.

```
> IsSymmetric_M(m1); IsAntiSymmetric_M(m1);
```
                              false
                              false

```
> IsSymmetric_M(m2); IsAntiSymmetric_M(m2);
```
                              true
                              false

```
>  IsSymmetric_M(m4); IsAntiSymmetric_M(m4);
```
$$false$$
$$true$$

6.5   Computing Closures of Relations

Determining the closures of a relation in Maple may be approached in
much the same way that we approached the problem of determining
properties of relations. Specifically, we shall implement algorithms that
use techniques similar to determining symmetric and reflexive properties
in order to determine the closure of these relational properties. The transitive closure of relation will require more insight, but we shall analyze
various methods for determining the transitive closure of a given relation.

Reflexive Closure

The algorithm for computing the reflexive closure of a relation is very
simple indeed. We simply set each diagonal entry in its matrix representation equal to 1. The resulting matrix represents the reflexive closure of
the relation.

```
>  RefClose := proc(M::matrix)
>    local i;
>    for i from 1 to coldim(M) do
>      M[i,i] := 1;
>    od;
>    evalm(M);
>  end:
```

We now use `RefClose` to find the reflexive closure of some of the relations
we have introduced as examples earlier in the chapter.

```
>  RefClose(m1); RefClose(m4);
```

$$\begin{bmatrix} 1 & 1 & 0 & 0 \\ 1 & 1 & 0 & 0 \\ 0 & 0 & 1 & 1 \\ 1 & 0 & 0 & 1 \end{bmatrix}$$

$$\begin{bmatrix} 1 & 0 & 0 & 0 \\ 1 & 1 & 0 & 0 \\ 1 & 1 & 1 & 0 \\ 1 & 1 & 1 & 1 \end{bmatrix}$$

Symmetric Closure

Next, we construct a procedure for determining the symmetric closure of a relation R. Again, we use the observation, from Page 382, which notes that, if (a, b) is an element of R, then (b, a) in an element of the symmetric closure of R. The Maple code, which is similar to the reflexive closure implemented above, follows.

```
>   SymmClose := proc(M::matrix)
>     local i, j;
>     for i from 1 to coldim(M) do
>       for j from 1 to rowdim(M) do
>         if M[i,j] = 1 then
>           M[j,i] := 1;
>         fi;
>       od;
>     od;
>     evalm(M);
>   end:
```

This procedure can be used to find the symmetric closures of some of our earler examples, as follows.

```
>   SymmClose(m1); SymmClose(m4);
```

$$\begin{bmatrix} 1 & 1 & 0 & 1 \\ 1 & 1 & 0 & 0 \\ 0 & 0 & 1 & 1 \\ 1 & 0 & 1 & 1 \end{bmatrix}$$
$$\begin{bmatrix} 1 & 1 & 1 & 1 \\ 1 & 1 & 1 & 1 \\ 1 & 1 & 1 & 1 \\ 1 & 1 & 1 & 1 \end{bmatrix}$$

Transitive Closure

Having created the simpler closures of the reflexive and symmetric properties, we now focus on implementing the transitive closure in Maple, which is a more difficult problem than the earlier cases in terms of computational complexity. In the text, there are two algorithms outlined, namely a generic transitive closure and Warshall's algorithm, and both will be covered in this section.

To implement the transitive closure, we need to implement both the Boolean join and the Boolean product operations that we previously introduced in Chapter 2. To begin, we will create the Boolean helper functions that enable us to convert between zero-one and true-false values.

```
>  with(linalg):
>  int_to_bool(0) := false:
>  int_to_bool(1) := true:
>  bool_to_int(true)  := 1:
>  bool_to_int(false) := 0:
```

Next, we construct the Boolean join function, again based on the previous work of Chapter 3.

```
>  BoolJoin := proc(A::matrix, B::matrix)
>    local i, j, C;
>    C := matrix(rowdim(A), coldim(A), zeroes);
>    for i from 1 to rowdim(A) do
>      for j from 1 to coldim(A) do
>        C[i,j] := int_to_bool(A[i,j]) or int_to_bool(B[i,j]);
>      od;
>    od;
>    map(bool_to_int,C);
>  end:
```

Following this, we construct the Boolean product.

```
>  BoolProd := proc(A::matrix, B::matrix)
>    local i, j, k, C;
>    C := matrix(rowdim(A), coldim(B), zeroes);
>    for i from 1 to rowdim(A) do
>      for j from 1 to coldim(B) do
>        C[i,j] := false;
>        for k from 1 to coldim(A) do
>          C[i,j] := C[i,j]
>                    or (int_to_bool(A[i,k])
>                        and int_to_bool(B[k,j]));
>        od;
>      od;
>    od;
>    map(bool_to_int, C);
>  end:
```

We are now ready to begin to implement the procedure for computing the transitive closure as defined on Page 387 of the text.

```
>  TransClosure := proc(M::matrix)
>    local i, A, B;
>    A := M;
>    B := A;
>    for i from 2 to coldim(M) do
>      A := BoolProd(A, M);
>      B := BoolJoin(B, A);
>      evalm(A);
>      evalm(B);
>    od;
```

```
>     evalm(B);
> end:
```

We test our transitive closure procedure on an example.

```
> T1 := matrix(3,3,[1,0,1,0,1,0,1,1,0]):
> TransClosure(T1);
```

$$\begin{bmatrix} 1 & 1 & 1 \\ 0 & 1 & 0 \\ 1 & 1 & 1 \end{bmatrix}$$

Next, we will examine how Warshall's algorithm compares (in terms of execution time on a simple example) to this general algorithm we have just implemented. First, we must implement Warshall's algorithm in Maple.

```
> Warshall := proc(M::matrix)
>    local i, j, k, W, n;
>    W := map(int_to_bool,M);
>    n := coldim(M);
>    for k from 1 to n do
>      for i from 1 to n do
>        for j from 1 to n do
>          W[i,j] := W[i,j] or (W[i,k] and W[k,j]);
>        od;
>      od;
>    od;
>    evalm(map(bool_to_int, W));
> end:
> Warshall(T1);
```

$$\begin{bmatrix} 1 & 1 & 1 \\ 0 & 1 & 0 \\ 1 & 1 & 1 \end{bmatrix}$$

We can compare these two procedures in terms of execution time using Maple's time command. But, we must note that this comparison on a single example does not prove anything; rather, it is useful for generally illustrating the execution times for the two algorithms that have been implemented. To do this illustration, we shall create a zero-one matrix that operates over the set $A = \{1, 2, 3, 4\}$.

```
> T2:=matrix(4, 4, [0,0,0,1,1,0,1,0,1,0,0,1,0,0,1,0]);
```

$$T2 := \begin{bmatrix} 0 & 0 & 0 & 1 \\ 1 & 0 & 1 & 0 \\ 1 & 0 & 0 & 1 \\ 0 & 0 & 1 & 0 \end{bmatrix}$$

```
>  st:=time():Warshall(T2):time()-st;
                        .033

>  st:=time():TransClosure(T2):time()-st;
                        .483
```

From this example, we can see that in Warshall's algorithm can be a substantial improvement over the method that uses Boolean joins and Boolean products, on this specific example. The reader is encouraged to explore this further.

6.6   Equivalence Relations

We shall examine, in this section, how we can use Maple to compute with equivalence relations. There are three specific problems that we shall address here: how to compute the equivalence class of an element, given an equivalence relation on some set; how to determine the number of equivalence relations on a finite set; and, how to compute the smallest equivalence relation that contains a given relation on some finite set.

To begin, we shall first provide a test for a relation to be an equivalence relation. Using the work that we have already done, and recalling that an equivalence relation is simply one that is reflexive, symmetric and transitive, our job is a simple one.

```
>  IsEqRel := IsTransitive @ IsSymmetric @ IsReflexive;
```
$$IsEqRel := IsTransitive@IsSymmetric@IsReflexive$$

Recall that, given an equivalence relation R, and a member a of the domain of R, the equivalence class of a is the set of all members b of the domain of R for which the pair (a, b) belongs to R. In other words, it is the set of all elements in the domain of R that are R-equivalent to a. So, the algorithm used to construct the equivalence class of a is very simple: we just search through R looking for all pairs of the form (a, b), adding each such second element b to the class. We do not have to search for pairs of the form (b, a), because equivalence relations are symmetric. Given an equivalence relation, and a point in its domain, this procedure returns the equivalence class of the point.

```
>  EqClass := proc(R::set, a::anything)
>    local i, S;
>    S := {};
>    for i from 1 to nops(R) do
>      if R[i][1] = a then
```

```
>         S := S union {R[i][2]};
>      fi;
>    od;
>    RETURN(S);
> end:
> EqClass({[0,0],[0,2],[1,0],[1,1],
>          [2,1],[1,2],[0,1]}, 0);
```
$$\{0, 1, 2\}$$

We now present a procedure that constructs all equivalence relations on a given set.

```
> DetermineEqClass := proc(A::set)
>   local P, Q, S, E, i, j, p;
>   S := {};
>   E := {};
>   for i from 1 to nops(A) do
>     for j from 1 to nops(A) do
>       S := S union {[A[i], A[j]]};
>     od;
>   od;
>   P := combinat[powerset](S);
>   for p in P do
>     if IsSymmetric(p)
>     and IsReflexive(p)
>     and IsTransitive(p) then
>       E := E union {p};
>     fi;
>   od;
>   RETURN(E);
> end:
> DetermineEqClass({1,2});
```
$$\{\{\}, \{[1, 1]\}, \{[2, 2]\}, \{[1, 1], [1, 2], [2, 2], [2, 1]\}, \{[1, 1], [2, 2]\}\}$$

```
> DetermineEqClass({1,2,3});
```
$$\{\{\}, \{[1, 1]\}, \{[2, 2]\}, \{[1, 1], [1, 2], [2, 2], [2, 1]\}, \{[1, 1], [2, 2]\},$$
$$\{[3, 3]\}, \{[1, 1], [1, 2], [2, 2], [3, 3], [2, 1], [3, 1], [3, 2], [1, 3], [2, 3]\},$$
$$\{[2, 2], [3, 3]\}, \{[1, 1], [3, 3]\}, \{[1, 1], [2, 2], [3, 3]\},$$
$$\{[1, 1], [2, 2], [3, 3], [3, 1], [3, 2], [1, 3], [2, 3]\},$$
$$\{[2, 2], [3, 3], [3, 2], [2, 3]\}, \{[1, 1], [2, 2], [3, 3], [3, 2], [2, 3]\},$$
$$\{[1, 1], [1, 2], [2, 2], [3, 3], [2, 1], [3, 2], [2, 3]\},$$
$$\{[1, 1], [1, 2], [2, 2], [3, 3], [2, 1]\}, \{[1, 1], [3, 3], [3, 1], [1, 3]\},$$
$$\{[1, 1], [2, 2], [3, 3], [3, 1], [1, 3]\},$$
$$\{[1, 1], [1, 2], [2, 2], [3, 3], [2, 1], [3, 1], [1, 3]\}\}$$

As the last question to be analyzed in this section, we shall determine the smallest equivalence relation containing a given relation. The motivating element in the algorithm is the fact that we need to generate a relation P containing the given relation R such that P is symmetric, reflexive and transitive. Recalling the section on closures, we deduce the following pseudocode algorithm for determining the equivalence relation P containing the relation R:

1. Create the reflexive closure of the relation R; call this P.
2. Create the symmetric closure of the relation P and call this Q. Note that Q is still reflexive since no elements were removed, so all the diagonal pairs (a, a) pairs still belong to Q.
3. Create the transitive closure of the relation Q and return this as output. This is reflexive for the same reason as outlined in the previous step. This relation is also symmetric since, if (a, b) and (b, c) entail the inclusion of an element (a, c), then since we executed the symmetric closure, we know that there are elements (c, b) and (b, a) so that we also have the element (c, a). Hence the final relation will be transitive, reflexive and symmetric.

We implement this in Maple as the composition of the four function `Warshall`, `SymmClose`, `RefClose`, and `MakeMatrix`.

```
>   EqContainment := Warshall @ SymmClose @ RefClose @ MakeMatrix;
```
$$EqContainment := Warshall@SymmClose@RefClose@MakeMatrix$$

```
>   R2;
```
$$\{[1, 1], [1, 2], [2, 1]\}$$

```
>   EqContainment(R2, {1,2,3,4});
```
$$\begin{bmatrix} 1 & 1 & 0 & 0 \\ 1 & 1 & 0 & 0 \\ 0 & 0 & 1 & 0 \\ 0 & 0 & 0 & 1 \end{bmatrix}$$

6.7 Partial Ordering and Minimal Elements

In this section of the textbook, we analyze *posets*, *maximal* and *minimal* elements, as well as the ideas of *least upper bounds*, *greatest lower bounds* and *topological sorting*. We will explore these topics in Maple, and will leave the exploration of the other topics of this section for the reader.

First, let us define a new Maple type for partial orders. For this section, we shall consider a partial order to be a set of ordered pairs (an object of type `rel`) that satisfies the three conditions necessary for a relation to be a partial order: reflexivity, antisymmetry and transitivity.

```
> 'type/po' := proc(obj)
>    type(obj, rel) and IsReflexive(obj)
>       and IsAntiSymmetric(obj)
>       and IsTransitive(obj);
> end:
```

This illustrates another way in which you can define new types in Maple, should the algebra of structured types prove to be inadequate.

Next, we shall construct a procedure that determines the set of minimal elements of a partially ordered set. The following procedure takes two arguments: a partial order R, and a subset S of the domain of R. It returns the set of minimal elements of S with respect to the ordering R.

```
> MinimalElements := proc(R::po, S::set)
>    local M,   # set of minimal elements of S; returned
>          s,t; # indices into S
>    if S minus DomainRelation(R) <> {} then
>      ERROR('set must be contained in the domain of the relation');
>    fi;
>    M := S;
>    for s in S do
>      for t in S minus {s} do
>        if member([t,s], R) then
>          M := M minus {s};
>        fi;
>      od;
>    od;
>    RETURN(M);
> end:
> R := DividesRelation({1,2,3,6});
```

$$R := \{[1,1], [1,2], [2,2], [3,3], [1,6], [2,6], [3,6], [6,6], [1,3]\}$$

```
> MinimalElements(R, {1,2,3,6});
```
$$\{1\}$$

```
> MinimalElements(R, {2,3,6});
```
$$\{2,3\}$$

Note that, by duality, we obtain – almost for free – a very simple implementation of MaximalElements.

```
> MaximalElements := proc(R::po, S::set)
>    MinimalElements(DualRelation(R), S);
> end:
> MaximalElements(R, {1,2,3,6});
```
$$\{6\}$$

```
> MaximalElements(R, {1,2,3});
```
$$\{2,3\}$$

Next, we shall construct a procedure for computing the least upper bound
of a set with respect to a given partial order, if it exists. Our procedure
will return the value NULL in case the set has no least upper bound. This
we shall accomplish in several steps. To do this, we first write a procedure
that will compute the set of all upper bounds of a subset of a partially
ordered set. This procedure, in turn, relies on the following utility for
determining whether a given element is an upper bound for a set.

```
>    IsUpperBound := proc(R::po, S::set, u::anything)
>      local s; # index into S
>      # sanity check
>      if not member(u, DomainRelation(R)) then
>        ERROR('bad arguments');
>      fi;
>      for s in S do
>        if not member([s, u], R) then
>          RETURN(false);
>        fi;
>      od;
>      RETURN(true);
>    end:
>    UpperBounds := proc(R::po, S::set)
>      local U,     # set of upper bounds of S; returned
>            DomR, # domain of R
>            d;     # index into DomR
>      DomR := DomainRelation(R);
>      # error checking
>      if S minus DomR <> {} then
>        ERROR('set must be contained in the domain of the relation');
>      fi;
>      U := {};
>      for d in DomR do
>        if IsUpperBound(R, S, d) then
>          U := U union {d};
>        fi;
>      od;
>      RETURN(U);
>    end:
```

Next, it is convenient to introduce a procedure mub that computes the set
of minimal upper bounds of a subset of a partially ordered set.

```
>    mub := proc(R::po, S::set)
>      MinimalElements(R, UpperBounds(R, S));
>    end:
```

Now, to complete the task at hand, we merely need to check whether mub
returns a singleton. If so, then the least upper bound exists (by definition);
otherwise, it does not, and we return the NULL value.

```
>    lub := proc(R::po, S::set)
>      local M; # set of minimal upper bounds of S
>      M := mub(R, S);
```

```
>    if nops(M) <> 1 then
>       RETURN(NULL);
>    fi;
>    RETURN(op(M));
>  end:
```

Topological sorting is used to produce, from a given partial order, a linear order on its domain that is compatible with it. For example, the natural order on the set $\{1, 2, 3, 6\}$ is a linear order that is compatible with the partial order of divisibility. (In fact, this is true of the lattice of divisors of any positive integer since, if m and n are positive integers, then m divides n only if $m \leq n$.) Having implemented least upper bounds and minimal elements, we can now create a topological sorting procedure that uses the above `MinimalElements` algorithm.

```
>  TopSort := proc(R::po, T::set)
>     local i, k, S, A;
>     k := 1;
>     S := T;
>     A := [];
>     while S <> {} do
>        A := [op(A), MinimalElements(R, S)[1]];
>        S := S minus {A[k]};
>        k := k+1;
>     od;
>     A;
>  end:
>  R := DivisorLattice(12);
```
$$R := \{[1, 1], [1, 2], [1, 4], [2, 2], [3, 3], [4, 4], [1, 6], [2, 6], [3, 6], [6, 6],$$
$$[12, 12], [1, 12], [2, 4], [3, 12], [2, 12], [4, 12], [1, 3], [6, 12]\}$$

```
>  TopSort(R, DomainRelation(R));
```
$$[1, 2, 3, 4, 6, 12]$$

```
>  R := DivisorLattice(2*3*5);
```
$$R := \{[1, 1], [1, 2], [2, 2], [3, 3], [1, 6], [2, 6], [3, 6], [6, 6], [1, 15],$$
$$[1, 30], [2, 10], [2, 30], [3, 15], [3, 30], [5, 5], [5, 10], [5, 15], [5, 30],$$
$$[1, 10], [1, 5], [6, 30], [10, 10], [10, 30], [15, 15], [15, 30], [30, 30], [1, 3]$$
$$\}$$

```
>  TopSort(DualRelation(R), DomainRelation(R));
```
$$[30, 6, 10, 2, 15, 3, 5, 1]$$

```
>  R := DivisorLattice(2*3*5*7);
```
$$R := \{[1, 1], [1, 2], [2, 2], [3, 3], [1, 6], [2, 6], [3, 6], [6, 6], [1, 15],$$
$$[1, 30], [2, 10], [2, 30], [3, 15], [3, 30], [5, 5], [5, 10], [5, 15], [5, 30],$$
$$[1, 10], [1, 5], [6, 30], [10, 10], [10, 30], [15, 15], [15, 30], [30, 30],$$
$$[1, 7], [1, 14], [1, 21], [1, 35], [1, 42], [1, 70], [1, 105], [1, 210], [2, 14],$$

$[2, 42], [2, 70], [2, 210], [3, 21], [3, 42], [3, 105], [3, 210], [5, 35],$
$[5, 70], [5, 105], [5, 210], [6, 42], [6, 210], [7, 7], [7, 14], [7, 21], [7, 35],$
$[7, 42], [7, 70], [7, 105], [7, 210], [10, 70], [10, 210], [1, 3], [15, 105],$
$[15, 210], [21, 21], [21, 42], [21, 105], [14, 14], [14, 42], [14, 70],$
$[14, 210], [30, 210], [35, 35], [35, 70], [35, 105], [35, 210], [42, 42],$
$[42, 210], [21, 210], [70, 70], [70, 210], [105, 105], [105, 210], [210, 210]$
$\}$

```
>  TopSort(R, numtheory[divisors](2*3*7));
```

$$[1, 2, 3, 6, 7, 14, 21, 42]$$

6.8   Lattices

In this section, we shall look at the problem of determining whether a partial order is a lattice. The approach that we shall take is a good example of "top down programming".

So that we may construct some interesting example, let's introduce a new function to produce examples of a nice class of lattices.

```
>  DivisorLattice := proc(n::posint)
>    DividesRelation(numtheory[divisors](n));
>  end:
```

The `divisors` procedure, from the `numtheory` package, returns the set of positive divisors of its integer argument. We use the procedure `DividesRelation`, constructed earlier in this chapter, to create the "divides" relation on this set.

```
>  type(DivRel(6), po);
```

true

We wish to write a Maple program that will determine whether a given (finite) partial order is a lattice. Now, a partial order R is a lattice if, and only if, it is both a *meet semilattice* and a *join semilattice*. The former is a partial order in which every pair of elements has a *meet* — a least upper bound, or supremum; the latter is one in which the dual condition is met: each pair of elements has a greatest lower bound, or infimum. So, our test for a partial order to be a lattice, merely needs to check that these two conditions are satisfied.

```
>  IsLattice := proc(R::po)
>    IsMeetSemiLattice(R) and IsJoinSemiLattice(R);
>  end:
```

Next, we use the fact that a partial order is a meet semilattice if, and only if, its dual relation is a join semilattice.

```
>  IsJoinSemiLattice := IsMeetSemiLattice @ DualRelation;
```
$$IsJoinSemiLattice := IsMeetSemiLattice@DualRelation$$

Now the real work begins; we must code the function `IsMeetSemiLattice`. For this, we must test whether, given a relation R, each pair a, b in the domain of R has a least upper bound with respect to R. One observation that simplifies our task considerably is that, since we are dealing only with finite relations, it suffices to check that each pair has a common upper bound.

```
>  IsMeetSemiLattice := proc(R::po)
>    local DomR,      # the domain of R
>          r,s;       # indices into R
>    DomR := DomainRelation(R);
>    for r in DomR do
>      for s in DomR do
>        if lub(R, {r, s}) = NULL then
>          RETURN(false);
>        fi;
>      od;
>    od;
>    RETURN(true);
>  end:
```

Finally, all the subroutines that go into making up our `IsLattice` program are complete, and we can test it on some examples. The following result should not come as any surprise.

```
>  IsLattice(DivisorLattice(24));
```
$$true$$

But, note what happens when we construct the "divides" relation on *all* the integers in the set $\{1, 2, 3, \ldots, 24\}$.

```
>  IsLattice(DividesRelation({seq(i, i=1..24)}));
```
$$false$$

6.9   Covering Relations

It is very inefficient, in general, to store all the pairs in a partial order relation. There is a minimal representation of a partial order, from which the entire relation can be reconstructed, called its *covering relation*. (This is not covered in the text, but will be useful to us in the following section, apart from being an important topic in its own right.) The covering

relation of a partial order is not itself a partial order. It is a minimal
subset of the partial order from which all the other relation pairs can be
deduced. Let R be a partial order on a set S. We say that an element
b in S *covers* an element a in S if (a, b) belongs to R, and $a \neq b$, but
there is no element $s \in S$ for which both (a, s) and (s, b) belong to R. In
other words, b covers a if b is greater than a, and if there is nothing in
between them. The *covering relation* of a partial order R is the relation C
on S consisting of those pairs (a, b) in R for which b covers a. As a simple
example, consider the set $\{1, 2, 3, 4\}$ ordered by magnitude. Its covering
relation is the set $\{(1, 2), (2, 3), (3, 4)\}$. All other pairs, such as $(1, 3)$, can
be "deduced" from the covering relation, using transitivity: $1 \leq 2 \leq 3$,
and hence, $1 \leq 3$ (that is, the pair $(1, 3)$ is in the relation). Covering
relations are also important because it is the covering relation of a partial
order that is drawn in a Hasse diagram, rather than the entire relation.

In this section, we provide a Maple procedure for computing the covering
relation of a partial order. First, we need a test for whether a given element
covers another.

```
>   Covers := proc(R::po, a, b)
>     local u;       # index into Dom(R)
>     if a = b then
>       RETURN(false);
>     fi;
>     if not member([a,b], R) then
>       RETURN(false);
>     fi;
>     for u in DomainRelation(R) minus {a, b} do
>       if member([a,u], R) and member([u,b], R) then
>         RETURN(false);
>       fi;
>     od;
>     RETURN(true);
>   end:
```

Now we can construct the covering relation of a partial order using the
following Maple procedure.

```
>   CoveringRelation := proc(R::po)
>     local C,       # covering relation; returned
>           DomR,    # the domain of R
>           r,s;     # indices into DomR
>     DomR := DomainRelation(R);
>     C := {};
>     for r in DomR do
>       for s in DomR do
>         if Covers(R, r, s) then
>           C := C union {[r,s]};
>         fi;
>       od;
```

```
>     od;
>     RETURN(C);
>   end:
```

Let's look at a few small examples.

```
>   CoveringRelation(DivisorLattice(6));
```
$$\{[1,2],[2,6],[3,6],[1,3]\}$$

```
>   CoveringRelation(DividesRelation({1,3,5,7,11,13,17}));
```
$$\{[1,5],[1,7],[1,17],[1,3],[1,11],[1,13]\}$$

6.10 Hasse Diagrams

The covering relation of a partial order of often used to represent a partial
order. We have already mentioned that it is more space efficient (requiring
less memory), and it is also used to represent a partial order graphically,
in the sense that the Hasse diagram is just a visual representation of the
covering relation of the partial order. We already have most of the tools
we need to make a first attempt at a visual representation of a partial
order. The visualization tools in the **networks** package make it relatively
easy to draw a graphical image of a partial order. We simply compute its
covering relation, and then convert it to a graph and display it as in the
following simple procedure.

```
>   HasseDiagramFirstTry := proc(R::po)
>     local C;
>     C := CoveringRelation(R);
>     networks[draw](MakeDigraph(DomainRelation(C),C));
>   end:
```

For instance, here is a picture of the divisor lattice of 210.

```
>   HasseDiagramFirstTry(DivisorLattice(2*3*5*7));
```

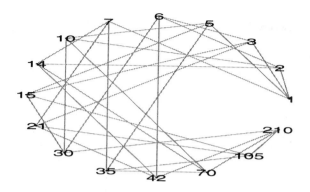

Regrettably, this suffers from the disadvantage of not drawing Hasse diagrams in the traditional way, with the order of the elements represented by the flow of the diagram. To fix this, we need to do a bit of coding. The idea is to arrange the elements of a partially ordered set into "levels", and then use the Linear option to the draw routine to render the diagram more appropriately. Several utility routines are needed. This function is used to check that an element is an *atom*; that is, that it has no predecessors. It is intended for use only with covering relations CR. The extra argument D is needed so that we can localize it later.

```
>   IsAtom := proc(CR::rel, D::set, a::anything)
>     local d;
>     for d in D do
>       if member([d,a], CR) then
>         # found a predecessor
>         RETURN(false);
>       fi;
>     od;
>     # must be an atom
>     RETURN(true);
>   end:
```

We use this in the next procedure, which determines all the atoms in a given subset of a covering relation.

```
>   Atoms := proc(CR::rel, D::set)
>     local A,   # set of atoms; returned
>           d;   # index into D
>     A := {};
>     for d in D do
>       if IsAtom(CR, D, d) then
>         A := A union {d};
>       fi;
>     od;
```

```
>     RETURN([op(A)]);
>   end:
```

Here is our new implementation of HasseDiagram. Most of the new work involves the arrangement of the elements of the partially ordered set into a sequence of "levels" to pass to Linear in the draw routine.

```
>  HasseDiagram := proc(R::po)
>    local L, C, G, A, D;
>    C := CoveringRelation(R);
>    D := DomainRelation(C); # = DomainRelation(R)
>    G := MakeDigraph(D, R);
>    L := NULL;
>    while D <> {} do
>      A := Atoms(C, D);
>      L := L, sort(A);
>      D := D minus {op(A)};
>    od;
>    networks[draw](Linear(L), G);
>  end:
```

It produces much nicer pictures, with the flow, from left to right, following roughly the increasing elements of the partially ordered set.

```
>  HasseDiagram(DivisorLattice(4));
```

```
>  HasseDiagram(DivisorLattice(6));
```

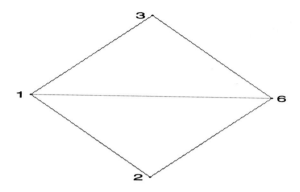

```
>   HasseDiagram(DivisorLattice(81));
```

The next two are especially beautiful!

```
>   HasseDiagram(DivisorLattice(2*3*5*7));
```

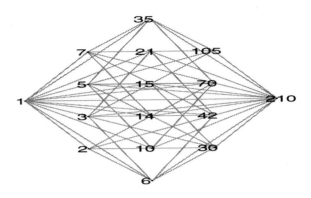

```
>   HasseDiagram(DivisorLattice(2*3*5*2));
```

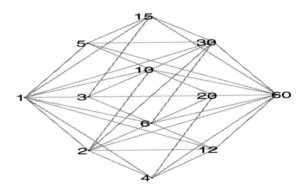

6.11 Computations and Explorations

In this section, we will explore how Maple can be used to solve questions 1, 4, and 5 of the *Computations and Explorations* section of the text.

1 Display all the different relations on a set with 4 elements.

Solution: As usual, Maple is much too powerful to solve only the single instance of the general problem suggested by this question. We provide here a very simple procedure that will compute all relations on any finite set.

```
>   AllRelations := proc(S::set)
```

```
>     local s, t,    # indices into S
>            C;        # Cartesian product SxS
>     C := {};
>     for s in S do
>       for t in S do
>          C := C union {[s,t]};
>       od;
>     od;
>     combinat[powerset](C);
>   end:
```

We now test our procedure on a smaller set, so as to keep the output to a reasonable length. The reader is encouraged to determine the running time and output length for the procedure when the input set has cardinality 4 or 5. Keep in mind that there are 2^{n^2} relations on a set with n members.

```
>   AllRelations({1,2});
```
$$\{\{\}, \{[1,1],[1,2],[2,1]\}, \{[1,1],[2,2],[2,1]\}, \{[1,1],[2,2]\},$$
$$\{[1,2],[2,1]\}, \{[2,2],[2,1]\}, \{[1,1],[2,1]\}, \{[1,1],[1,2],[2,2]\},$$
$$\{[1,2],[2,2]\}, \{[1,2],[2,2],[2,1]\}, \{[2,2]\},$$
$$\{[1,1],[1,2],[2,2],[2,1]\}, \{[1,2]\}, \{[1,1],[1,2]\}, \{[2,1]\}, \{[1,1]\}\}$$

4 Determine how many transitive relations there are on a set with n elements for all positive integers n with $n \leq 7$.

Solution: We will construct each possible $n \times n$ zero-one matrix using an algorithm similar to binary counting. The pseudocode is:

1. For each value from 1 to $2^{(n^2)}$, we create a list that is the base 2 representation of that integer.

2. We create a matrix M that has elements being that list of values. These are all possible n^2 zero-one matrices (the reader can prove this statement).

3. Evaluate the transitive closure of each of the matrices we have created, and return the set of these transitive closure matrices.

The implementation is as follows;

```
>   FindTransitive := proc(S::set)
>     local i, j, T, P;
>     P := {};
>     for i from 0 to 2^(nops(S)^2)-1 do
>       T := convert(i, base, 2);
>       while nops(T) < nops(S)^2 do
>          T := [op(T), 0];
>       od;
>       P := P union {matrix(nops(S), nops(S), T)};
>     od;
>     P;
>   end:
```

Again, we use our procedure on relatively small values (owing to the length of the output), and leave further exploration to the reader.

```
>   P := FindTransitive({1,2}):
>   Q := {}:
>   for i from 1 to nops(P) do
>      Q := Q union {Warshall(P[i])}:
>   od:
>   Q;
```

$$\left\{ \begin{bmatrix} 1 & 1 \\ 1 & 1 \end{bmatrix}, \begin{bmatrix} 1 & 0 \\ 0 & 0 \end{bmatrix}, \begin{bmatrix} 1 & 0 \\ 1 & 1 \end{bmatrix}, \begin{bmatrix} 0 & 0 \\ 1 & 0 \end{bmatrix}, \begin{bmatrix} 0 & 1 \\ 0 & 1 \end{bmatrix}, \begin{bmatrix} 0 & 1 \\ 0 & 0 \end{bmatrix}, \begin{bmatrix} 0 & 0 \\ 0 & 0 \end{bmatrix}, \begin{bmatrix} 1 & 0 \\ 1 & 0 \end{bmatrix}, \begin{bmatrix} 1 & 1 \\ 0 & 1 \end{bmatrix}, \right.$$

$$\left. \begin{bmatrix} 1 & 0 \\ 0 & 1 \end{bmatrix}, \begin{bmatrix} 0 & 0 \\ 0 & 1 \end{bmatrix}, \begin{bmatrix} 1 & 1 \\ 0 & 0 \end{bmatrix}, \begin{bmatrix} 0 & 0 \\ 1 & 1 \end{bmatrix} \right\}$$

5 Find the transitive closure of a relation on a set with at least 20 elements.

Solution: We will generate a random zero-one matrix with dimension 10×10, and then apply Warshall's algorithm to deduce the matrix's transitive closure.

To generate a random zero-one matrix, we use the `randmatrix` function from the `linalg` package, and supply its third, optional argument with a procedure that generates a random sequence of 0s and 1s. We then apply Warshall's algorithm to this random matrix, obtaining the result. Here, to conserve space, we apply this procedure to a 10×10 matrix as an example.

```
>     Q := randmatrix(10, 10, entries=rand(2));
```

$$Q := \begin{bmatrix} 1 & 0 & 1 & 1 & 0 & 0 & 1 & 0 & 1 & 1 \\ 1 & 1 & 1 & 0 & 0 & 1 & 0 & 1 & 1 & 1 \\ 0 & 0 & 0 & 1 & 1 & 1 & 0 & 0 & 1 & 0 \\ 1 & 1 & 0 & 1 & 0 & 1 & 0 & 0 & 1 & 1 \\ 1 & 0 & 1 & 0 & 0 & 0 & 1 & 1 & 1 & 1 \\ 1 & 0 & 1 & 1 & 1 & 1 & 1 & 1 & 1 & 1 \\ 0 & 0 & 1 & 1 & 1 & 1 & 0 & 1 & 0 & 0 \\ 1 & 0 & 1 & 1 & 0 & 1 & 1 & 0 & 1 & 0 \\ 0 & 1 & 1 & 0 & 0 & 0 & 1 & 0 & 1 & 0 \\ 1 & 1 & 1 & 1 & 1 & 1 & 0 & 1 & 0 & 1 \end{bmatrix}$$

```
>    Warshall(Q);
```

$$\begin{bmatrix}
1 & 1 & 1 & 1 & 1 & 1 & 1 & 1 & 1 & 1 \\
1 & 1 & 1 & 1 & 1 & 1 & 1 & 1 & 1 & 1 \\
1 & 1 & 1 & 1 & 1 & 1 & 1 & 1 & 1 & 1 \\
1 & 1 & 1 & 1 & 1 & 1 & 1 & 1 & 1 & 1 \\
1 & 1 & 1 & 1 & 1 & 1 & 1 & 1 & 1 & 1 \\
1 & 1 & 1 & 1 & 1 & 1 & 1 & 1 & 1 & 1 \\
1 & 1 & 1 & 1 & 1 & 1 & 1 & 1 & 1 & 1 \\
1 & 1 & 1 & 1 & 1 & 1 & 1 & 1 & 1 & 1 \\
1 & 1 & 1 & 1 & 1 & 1 & 1 & 1 & 1 & 1 \\
1 & 1 & 1 & 1 & 1 & 1 & 1 & 1 & 1 & 1
\end{bmatrix}$$

The reader can explore larger examples.

6.12   Exercises/Projects

Exercise 1. Write a Maple procedure with the signature

$$\texttt{mkRelation(S:set(integer), e:expression)}$$

that creates the ordered pair relation $\{(a, b) \in S \times S : e(a, b) \text{ is true}\}$. That is, mkRelation should return the set of all ordered pairs (a, b) of integers for which the expression evaluates to **true** when a and b are substituted for the variables in e. The input expression e should be a boolean valued Maple expression involving two integer variables x and y as well as operators that take integer operands. For example, your procedure should accept an expression such as x + y < x * y.

Exercise 2. Write a Maple procedure to generate a random relation on a given finite set of integers.

Exercise 3. Use the procedure you wrote in the preceding question to investigate the probability that an arbitrary relation has each of the following properties: (a) reflexivity; (b) symmetry; (c) anti-symmetry; and (d) transitivity.

Exercise 4. Write Maple procedures to determine whether a given relation is irreflexive or asymmetric. (See the text for definitions of these properties.)

Exercise 5. Define new Maple types for equivalence relations and for lattices.

Exercise 6. Investigate the ratio of the size of an arbitrary relation to the size of its transitive closure. How much does the transitive closure make a relation "grow", on average?

Exercise 7. Examine the function φ defined as follows. For a positive integer n, we define $\varphi(n)$ to be the number of relations on a set of n elements whose transitive closure is the "all" relation. (If A is a set, then the "all" relation on A is the relation $A \times A$ with respect to which every member of A is related to every other member of A, as well as to itself.)

Exercise 8. Write a Maple procedure that finds the antichain with the greatest number of elements in a partial ordering. (See Page 424 of the text for definitions.)

Exercise 9. The **transitive reduction** of a relation G is the smallest relation H such that the transitive closure of H is the transitive closure of G. Use Maple to generate some random relations on a set with ten elements and find the transitive reduction of each of these random relations.

Exercise 10. Write a Maple procedure that computes a partial order, given its covering relation.

Exercise 11. Write a maple procedure to determine whether a given lattice is a Boolean algebra, by checking whether it is distributive and complemented. (See Page 615 of the text for definitions.)

7 Graphs

This chapter covers material related to graph theory. In particular, we describe how to do computations with graphs using Maple. Working with graphs computationally requires techniques for representing graphs, displaying graphs, manipulating graphs, and using algorithms to solve particular types of problems involving graphs. Maple directly supports all of these operations. We explain how to input graphs in Maple, how to display graphs with Maple, how to alter graphs with Maple, and how to solve a wide variety of graph problems, using many of the algorithms that are already implemented as part of the Maple system. We show how to use Maple to determine whether a graph is connected, whether it is bipartite, whether it has an Euler circuit, whether it has an Hamilton circuit, whether it is planar, and so on. We also show how to use Maple to find the shortest path between vertices in a weighted graph.

7.1 Getting Started with Graphs

Simple Graphs

The first thing we need to do in order to work with graphs using Maple is to learn how to create them. We begin with simple graphs. Recall that a simple graph, as defined on Page 430 of the text, is a set V of vertices, and a set E of unordered pairs of elements of V, called edges of the graph.

Maplehas a large collection of commands related to graph theory, all found in the **networks** package. In order to access these commands we use the **with** command to load in the entire package. This action defines the various graph specific procedures that are available in Maple. A list of the names of the actual commands that become available in this way appears in response to executing the command.

```
>  with(networks);
```
   [*acycpoly, addedge, addvertex, adjacency, allpairs, ancestor, arrivals,*
       bicomponents, charpoly, chrompoly, complement, complete, components,
       connect, connectivity, contract, countcuts, counttrees, cube, cycle,
       cyclebase, daughter, degreeseq, delete, departures, diameter, dinic,
       djspantree, dodecahedron, draw, duplicate, edges, ends, eweight,
       findroot, flow, flowpoly, fundcyc, getlabel, girth, graph, graphical,
       gsimp, gunion, head, icosahedron, incidence, incident, indegree, induce,
       isplanar, maxdegree, mincut, mindegree, neighbors, new, octahedron,
       outdegree, path, petersen, random, rank, rankpoly, shortpathtree, show,

shrink, span, spanpoly, spantree, tail, tetrahedron, tuttepoly, undirected,
vdegree, version, vertices, void, vweight]

Once we have loaded the package in this manner, we can use the newly
defined Maple commands to create new graphs and then add or delete
edges and vertices, or even contract edges. Subsets of the vertices can
be used to induce subgraphs. Some of the commands construct directly
various special types of graphs such as complete graphs, hyper-cubes,
the Petersen graph, and random graphs. Other commands compute some
of the important characteristics of a given graph, such as its maximum
degree, its diameter or its planarity. In Chapter 8, we will explore in
more detail some of the tree specific functions of this package such as
`djspantree` for determining spanning trees of a graph and `daughter` for
identifying the descendents of a node in a directed graph.

To create a new graph called `G1`, we use the `new` command. This command to creates a new graph "object" and initializes all the underlying
structures of the graph. The command only accepts one argument — the
name of the new graph.

```
>  G1 := new():
```

This newly created graph does not yet have any vertices or edges. These
must be added to the graph using such commands as `addvertex`, `addedge`,
and `connect`.

We continue to construct an example of a simple graph by adding vertices
to represent cities and edges to represent the communications links of
a computer network between the chosen cities. We use the `addvertex`
command to add vertices and the `addedge` command to add edges and
build up the graph.

For our example we have chosen a set of seven American cities – Los
Angeles, Denver, Detroit, Washington, New York and San Francisco. The
following command adds this set of seven vertices to the newly created
graph `G1`.

```
>  addvertex( {'Los Angeles', Denver, Chicago, Detroit,
>             Washington, 'New York', 'San Francisco'},
>     G1);
```
New York, Detroit, Washington, San Francisco, Los Angeles, Chicago,
  Denver

The newly created vertex names are listed in response to the command.

To complete our graph representing a nation-wide computer network we
next add the edges representing the connections between the various cities.

Each connection is undirected. Information may flow in either direction. Undirected connections are defined by using a set of two vertices – the two cities joined by the connection. To add such edges to our graph we use the **addedge** command in a similar manner to the use of the **addvertex** command. This time the first argument to the procedure call is the set of vertex pairs representing the new edges. The last argument is the name of the graph to which the edges are to be added.

```
>   addedge(
>     {{'Los Angeles', Denver}, {'Los Angeles', 'San Francisco'},
>      {'San Francisco', Denver}, {Denver, Chicago},
>      {Chicago, Detroit}, {Chicago, Washington},
>      {Chicago,'New York'}, {Detroit,'New York'},
>      {Washington, 'New York'}},
>   G1);
```

$$e1, e2, e3, e4, e5, e6, e7, e8, e9$$

As each edge is created, a name is generated. The edge names are displayed as the value of this operation. The names are used so that, if necessary, more than one connection can occur between cities.

To discover the vertex pair associated with a particular edge in **G1**, use the **ends** command, as in

```
>   ends(e1,G1);
```

$$\{New\ York, Detroit\}$$

To recover the set of all pairs corresponding to edges of a particular graph use the command

```
>   ends(G1);
```

$$\{\{New\ York, Detroit\}, \{Los\ Angeles, Denver\},$$
$$\{San\ Francisco, Los\ Angeles\}, \{San\ Francisco, Denver\},$$
$$\{Chicago, Denver\}, \{Detroit, Chicago\}, \{Washington, Chicago\},$$
$$\{New\ York, Chicago\}, \{New\ York, Washington\}\}$$

Visualizing Graphs in Maple

The usefulness of graphs is realized partly through our ability to draw diagrams representing graphs. Such visual presentations of graphs sometimes lead to a clearer understanding of the underlying relationships represented by the graphs. The beauty of some of the resulting diagrams is also one of the things that helps to make this such a popular subject.

In Maple, we present a graph visually by using the **draw** command. There are usually many pictorial presentations possible for a given graph, depending on how the vertices are placed and how they are connected. In

the absence of additional instructions the draw command simply places
the vertices of the graph counter-clockwise around a circle and connects
the appropriate vertices by straight lines, as in

> draw(G1);

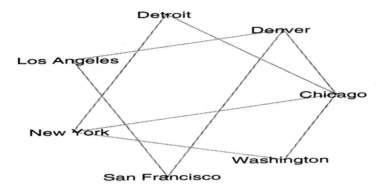

Later we will show how to use the draw command with particular instructions to draw graphs in a variety of ways. We see from the diagram that
some of the communications links cannot ever be allowed to fail if all the
cities are to remain connected. To improve reliability of the network we
may wish to add additional edges. For example, to add an edge between
Denver and Chicago, use addedge again as in

> addedge({Denver,Washington},G1);

$$e10$$

Note that this produces a new edge called e10. To see the results use draw
again.

> draw(G1);

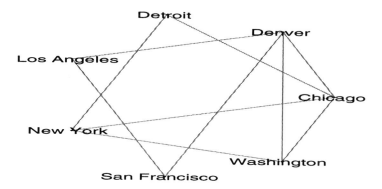

Loops and Multiple Edges

Graphs can have loops and multiple edges. For example, we may wish to add multiple edges between Denver and Detroit to improve reliability over that portion of the network. The command

```
>   addedge({Denver,Detroit},G1);
```
$$e11$$

adds a new edge `e11` which is different from the earlier. To add a loop at New York, use the command

```
>   addedge( {'New York', 'New York' }, G1);
```
$$e12$$

Unfortunately, due to current limitations of the `draw` command, such additional edges do not show up on the diagrams that are generated by Maple.

Note also, that the degree command ignores loops. That is,

```
>   degree( 'New York' , G1);
```
$$0$$

```
>   edges( {'New York'}, G1);
```
$$\{e12\}$$

This fact must be taken into account carefully when looking at results involving degrees on non-simple graphs.

It is helpful to begin by ensuring that your graph is simple. The `gsimp` command can be used to achieve this by stripping off loops and collapse multiple edges into a single edge.

```
>  gsimp( G1 ):
>  edges( {'New York'}, G1);
```

$$\{\}$$

The result is a simple graph.

Alternative Presentations

The drawing algorithm based on positioning vertices around a circle is surprisingly effective, but there is still a need for alternative presentations. The right visual presentation can be very effective at revealing the underlying structure of the underlying graph.

The `networks` package is designed in such a way that every graph can supply its own drawing routine. Detailed specifications of the layout can be provided and saved as part of the graph object. An example of such a specification appears in the section on page 235.

Detailed layouts of particular graphs can be extremely tedious to construct. To help avoid this tedium, Maple provides several additional methods for specifying graphic layouts. The basic `draw` routines has access to two additional general methods of determining a graphical layout. One of them, `Linear`, is most appropriate for drawing graphs in which the edges tend to go between different groups of vertices and the other, `Concentric`, is appropriate for graphs built out of nested cycles. Using them, here are two other ways to draw the complete graph K_8, on 8 vertices.

```
>  K8 := complete(8):
```

The default presentation is:

```
>  draw(K8);
```

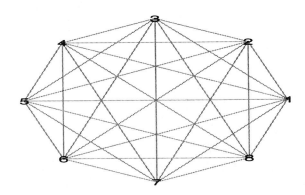

Use of the `Linear` option arranges the vertices of the graph to be displayed
so that the lists of vertices are placed in linear sequences, as in

```
> draw(Linear([1,3,5,7], [2,4,6,8]), K8);
```

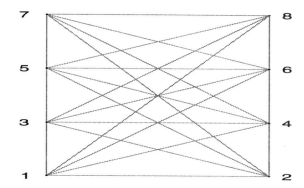

The `Concentric` option arranges the vertices in concentric circles, with
the vertices in each list placed in its own circle in the diagram.

```
> draw(Concentric([1,2,3,4], [5,6,7,8]), K8);
```

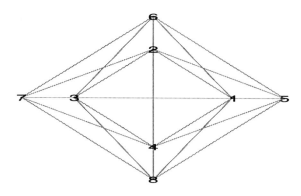

The precise appearance of the graph depends on how the vertices are
arranged into linear lists or circles.

```
> draw(Linear([1,2,3,4], [5,6,7,8]), K8);
```

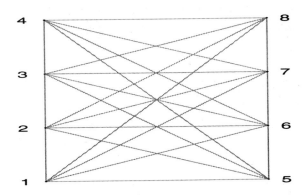

```
>   draw(Concentric([1,3,5,7], [2,4,6,8]), K8);
```

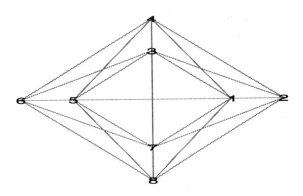

Note how the labeling has changed from the two earlier examples. Of course, there can be several levels of concentric circles, or several linear arrangements in a graph diagram.

```
>   K9:= complete(9):
>   draw(Linear([1,2,3,4,5],[6,7,8],[9]), K9);
```

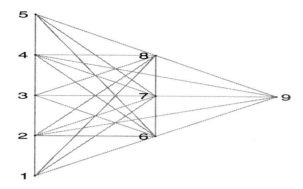

Here is an especially intricate example.

```
>   K25 := complete(25):
>   draw(Concentric( [1,2,3,4,5],
>    [6,7,8,9,10], [11,12,13,14,15],
>    [16,17,18,19,20], [21,22,23,24,25]
>   ), K25);
```

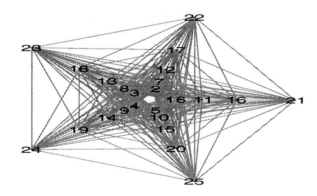

You will need a display with a fairly high resolution to see the details of the complex picture.

Directed Graphs

Next, we show how to represent directed graphs in Maple. To do this, we represent edges as *ordered pairs* of two vertices rather than as *sets* of two vertices. For example, if we add one more alteration to our computer network, where we are limited to only unidirectional lines of transmission,

we would need to create a directed graph in order to represent this new
computer network. This can be easily accomplished by signifying an edge
as an ordered list of 2 vertices instead of an (unordered) set of 2 vertices.
As an example, consider the graph G2 which has the same vertex set as
graph G1, but has only unidirectional information lines between cities. We
would represent such a computer network as the following directed graph
in Maple:

```
>   G2 := new():
>   addvertex( {'Los Angeles', Denver, Chicago, Detroit,
>       Washington, 'New York', 'San Francisco'},
>   G2);
```
New York, Detroit, Washington, San Francisco, Los Angeles, Chicago,
  Denver

```
>   addedge(
>   {['Los Angeles', Denver], ['Los Angeles', 'San Francisco'],
>    ['San Francisco', Denver], [Denver, Chicago],
>    [Chicago, Detroit], [Chicago, Washington],
>    [Chicago, 'New York'], ['New York', Detroit],
>    [Washington, 'New York']},
>   G2);
```
$$e1, e2, e3, e4, e5, e6, e7, e8, e9$$

```
>   draw(G2);
```

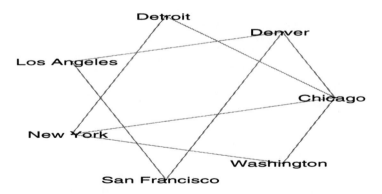

Notice that loops and multiple directed edges can be constructed in Maple
in the same manner that undirected multiple edges and loops were added.
Specifically,

```
>   addedge( {[Denver, Denver], [Chicago, Washington],
>       [Washington,Chicago], [Washington, Chicago]},
>   G2):
>   draw(G2);
```

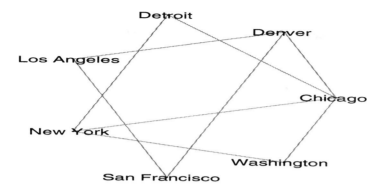

Next, we will show how to delete edges and vertices and how to contract edges. The motivation for the deletion of an edge (or vertex) in a graph is the scenario that a particular phone line (or city center) of a computer network becomes inoperable due to technical difficulties. In Maple, we can delete an edge, or a vertex, by using the **delete** command in the following manner:

```
>   delete(edges(['Los Angeles', Denver], G2),G2):
>   draw(G2);
```

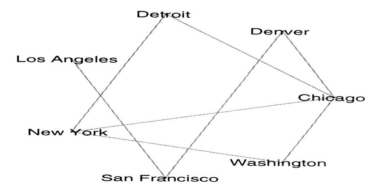

From the pictorial representation, we notice that there is now no edge connecting Denver to Los Angeles. If there was a case in which a city was no longer able to connect to the computer network represented by graph G2, we could use the **delete** command, with a vertex as the first parameter. Specifically, if for some reason, the entire city of Chicago was no longer attached to the computer network, due to a natural disaster, say, then we can represent this new network as:

```
>  delete(Chicago, G2):
>  draw(G2);
```

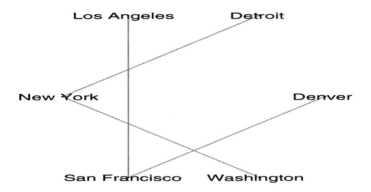

As a last example of how the computer network may be altered, we consider the case where two cities merge their computational resources into a single site. In this case, we would use the Maple `contract` command, to contract an edge, which can be represented as two vertices or a single edge. Specifically, let us merge the computer sites of San Francisco and Denver into a single node.

First we must identify the edge between the two cities. In general, there may be several edges going each direction as well as bi-directional edges. To determine which edges are present use one of the following variations of the `edges` command.

The edges directed from San Francisco to Denver are given by the commands

```
>  edges( ['San Francisco' , 'Denver' ] , G2 );
```
$$\{e4\}$$

The edges directed from Denver to San Francisco are:
```
>  edges( [ 'Denver', 'San Francisco' ] , G2 );
```
$$\{\}$$

The complete collection of all edges between the two cities is given by
```
>  edges( { 'Denver', 'San Francisco' } , G2 , 'all' );
```
$$\{e4\}$$

We can describe the edge to be contracted by their names such as `e1`, or implicitly by using lists or sets of vertices. For example, the following command accomplishes the contraction.

```
> contract(['San Francisco', Denver ], G2):
> draw(G2);
```

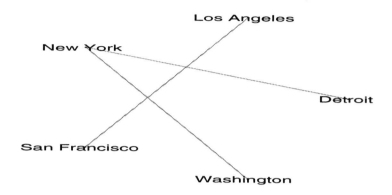

Notice that there is a new loop introduced at the `Chicago` vertex, resulting from the contraction of the edge joining `San Francisco` to `Denver`.

```
> edges( {'San Franciso', 'San Francisco'} , G2 , 'all' );
```
$$\{\}$$

Again, Maple's current `draw` command makes no attempt to handle direction, loops or multiple edges so it is not visible on the graph.

In general, Maple can be used to construct graphs with a mixture of directed or undirected edges, with or without multiple edges, and with or without loops all in one graph. None of this structure is discarded internally, but the orientation of an edge set or the multiplicity of edges and the presence of loops are not reflected in Maple's diagrams of graphs. All graphs are displayed by Maple as simple graphs, even if they may actually have a structure that violates simplicity. Later in this chapter, we will show how to represent weighted graphs, that is, graphs for which weights are assigned to edges.

7.2   Simple Computations on Graphs

Having shown how to use Maple to construct and draw various types of graphs, we now show how to use Maple to determine some graph parameters. In doing so, we can begin to determine similarities and differences between graphs as well as deriving properties of a graph, such as that of containing an Eulerian path (which will be covered later in this chapter) if certain vertex degree conditions are met.

Degrees of Vertices

To begin, we shall show how to find the degree of a vertex in an undirected graph, using the `vdegree` command. Using our graph G1, defined earlier, which represents the computer network across several cities, we can determine the number of lines into a specific city, which is simply the degree of the vertex that represents that city in our simple graph.

```
> vdegree(Detroit, G1);
```
$$3$$

```
> vdegree('San Francisco', G1);
```
$$2$$

We can create a simple procedure, which we shall call `DegCount` to sum the degrees of all the vertices, and then count the number of edges. Comparing these two quantities will verify (but not prove) the Handshaking Theorem as stated on Page 438 of the text.

```
> DegCount := proc(G::graph)
>    local V, v, deg_sum;
>    V := vertices(G);
>    deg_sum := 0;
>    for v in V do
>      deg_sum := deg_sum + vdegree(v, G);
>    od;
>    printf('The sum of the degrees is %d\n', deg_sum);
>    printf('The number of edges is %d\n', nops(edges(G)));
>    end:
> DegCount(G1);

The sum of the degrees is 22
The number of edges is 11
```

We shall extend this procedure to apply a similar technique to *directed* graphs. First, we note that Maple has commands for finding the in-degree and out-degree of vertices in directed graphs. These are the `indegree` and `outdegree` commands, respectively.

We illustrate the use of these commands with the example of the directed graph representing the computer network, namely G2.

```
> outdegree('New York', G2);
```
$$1$$

```
> indegree('New York', G2);
```
$$1$$

Maple also has functions that determine the sets of vertices of a directed graph that are adjacent to a vertex by outgoing and incoming edges. These functions are called **departures** and **arrivals**, respectively. We illustrate their use by providing another way to find the outdegree and indegree of a vertex, namely by counting the number of departures and the number of arrivals. We use our previous example.

```
> arrivals('New York', G2);
```
$$\{Washington\}$$

```
> nops(arrivals('New York', G2));
```
$$1$$

```
> nops(departures('New York', G2));
```
$$1$$

We will now implement a procedure, called **DirectedDegCount**, which is the directed graph version of **DegCount**, which was used for undirected graphs.

```
> DirectedDegCount := proc(G::graph)
>    local V, v, in_sum, out_sum;
>    V := vertices(G);
>    in_sum := 0;
>    out_sum := 0;
>    for v in V do
>       in_sum := in_sum + indegree(v, G);
>       out_sum := out_sum + outdegree(v, G);
>    od;
>    printf('The sum of the outdegrees is %d\n', out_sum);
>    printf('The sum of the indegrees is %d\n', in_sum);
>    printf('The number of edges in the given digraph is %d\n',
>           nops(edges(G)));
> end:
```

We now use this procedure to verify Theorem 3 for the particular graph G2 as follows:

```
> DirectedDegCount(G2);
```

```
The sum of the outdegrees is 3
The sum of the indegrees is 3
The number of edges in the given digraph is 4
```

7.3   Constructing Special Graphs

We now show how to use Maple to work with certain special types of graphs. There are many families of graphs that occur commonly in applications, and Maple provides direct support for working with graphs from these families.

To begin, we shall explore complete graphs.

A complete graph is a simple, undirected graph that has all possible edges. Since there is an edge between every pair of vertices, there are $n(n-1)/2$ edges in a complete graph with n vertices. Maple has a function, **complete**, for generating complete graphs. For example, we can generate, and display, the complete graphs K_5 and K_7, on 5 and 7 vertices respectively, as follows:

```
> K5:=complete(5):
> draw(K5);
```

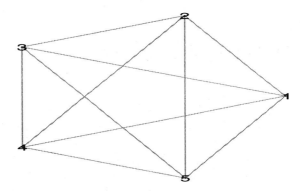

```
> K7:=complete(7):
> draw(K7);
```

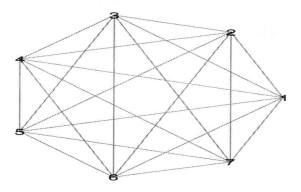

Notice that K_5 and K_7 are drawn by Maple as a regular pentagon and septagon, with lines connecting each vertex.

In a similar manner to constructing complete graphs, we may construct cycles, which are simple undirected graphs that are connected and have n edges if there are n vertices. As examples, we will construct cycles on 6 and 9 vertices, which are denoted as C_6 and C_9 respectively;

```
>   draw(cycle(6));
```

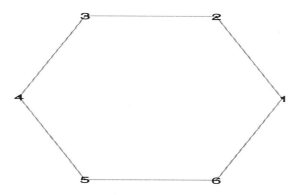

```
>   draw(cycle(9));
```

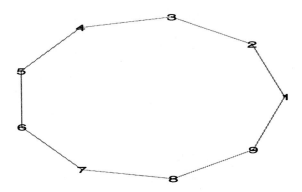

Note that the Maple pictorial representation of the cycle on n vertices is precisely a regular n-gon.

Similarly, we may construct the n-cube on 2^n vertices, as outlined on Page 441 of the text, using the **cube** command; For example, we draw the 2-cube and the 3-cube.

```
>  draw(cube(2));
```

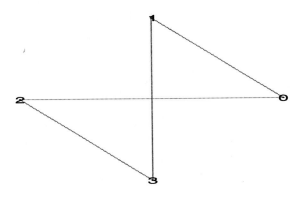

```
>  draw(cube(3));
```

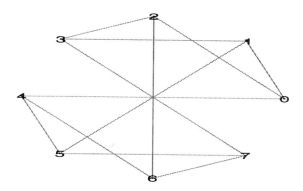

Each n-cube is constructed from two $n-1$-gon's, by connecting the vertices pairwise. There are 2^n vertices in total.

Maple also has commands for constructing certain important graphs, including the Petersen graph and graphs formed by the vertices and edges of regular polyhedra.

First, we will examine how to construct a Petersen graph. The Petersen graph is a graph on 10 vertices such that each vertex has degree 3, and is introduced on Page 487 of the text. Maple has the `petersen` procedure, that creates a Petersen graph.

```
>  draw(petersen());
```

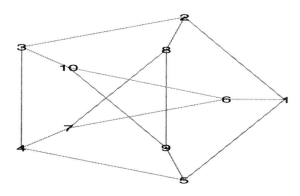

Using Maple, we may also draw the vertices and edges of several regular polyhedra. Specifically, Maple can be used to construct tetrahedra, octahedra, dodecahedra, and icosahedra, which are regular polyhedra with

4, 8, 12, 20 sides, respectively. This means that these are 3-dimensional objects that have sides that are regular polygons, such as pentagons or triangles. When we consider the corner points of these shapes as vertices, and taking the intersection of two sides of the polyhedron as edges, we can express these polyhedra as graphs. For instance, we can use Maple to draw a graph representing a dodecahedron using the `dodecahedron` command as follows.

```
> draw(dodecahedron());
```

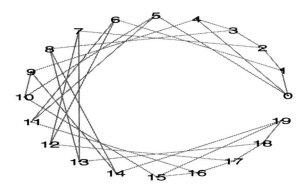

Additionally, we can draw the graph representing a 12-sided regular polyhedron using the Maple `icosahedron` command similar to the function above.

```
> draw(icosahedron());
```

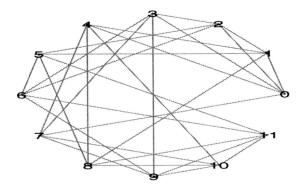

The reader can examine the other two functions (`octahedron` and `tetrahedron`) that generate polyhedra.

Bipartite Graphs

Another important class of graphs is the class of bipartite graphs. A bipartite graph is one whose vertex set can be partitioned into two disjoint sets such that any edge has exactly one vertex in each partitioning set.

First, we will show how to construct the complete bipartite graphs. To do this, we make use of the `complete` function, covered earlier in this chapter. By definition, a complete bipartite graph is a bipartite graph with bipartition $V = (A, B)$ such that there are m vertices in A and n vertices in B, and there is an edge joining every vertex in A to every vertex in B. Thus, we can see that there are mn edges in total in the complete bipartite graph $K_{m,n}$. We can use Maple to create a $K_{m,n}$ on any combination of positive integers m and n. For example, we can create the complete bipartite graphs $K_{3,3}$ and $K_{2,4}$.

```
>  draw(complete(3,3));
```

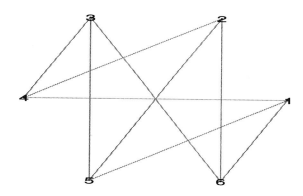

```
>  draw(complete(2,4));
```

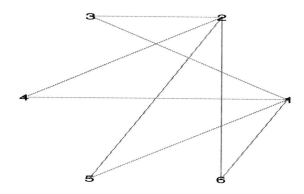

The `complete` command can construct complete n-partite graphs, for any positive integer n (provided, of course, that the resulting object will fit into the computer's memory).

It is worth noting that Maple attempts to draw the bipartite graphs with the one vertex partition on one side of a regular $m + n$-gon, and the other vertex partition on the other side of the $m + n$-gon.

Using Maple, we can construct a procedure that will determine whether a graph is bipartite or not, and return a vertex partition if the graph is bipartite. (Note: the method used here is based on forming a spanning tree of the graph, a topic covered in depth in Chapter 8 of the text).

A description of the procedure, which will be called `Bipartite`, is as follows.

1. Pick a vertex v from the vertex set. Place it in set A.

2. Place all of v's neighbors in set B.

3. Pick each unchosen vertex, w, of B. Place all of the neighbors of w that are neither in A nor in B into set A.

4. Repeat step 3, reversing A and B, until there are no more unchosen vertices.

5. At this stage, we have formed a disjoint partition of the vertices (the reader can verify this). We now examine each edge of the graph and ensure that no edge has both ends in A nor both ends in B. We return the result of our search, as well as the partition of vertices if our edge verification came back true.

To begin, we will need to the use of a simple procedure that generates the set product of a set with itself.

```
>   SetProd:=proc(S::set)
>     local i, j, R;
>     R:={};
>     for i from 1 to nops(S) do
>       for j from 1 to nops(S) do
>         R:=R union {{S[i], S[j]}};
>       od;
>     od;
>     R;
>   end:
```

As an example of the usage of this function, consider

```
>   SetProd({1,2,3});
```
$$\{\{1\}, \{1,2\}, \{1,3\}, \{2,3\}, \{3\}, \{2\}\}$$

The implementation of the pseudocode for `Bipartite` is as follows;

```
>   Bipartite := proc(G::graph)
>     local i, j, v, w, AB, R, E, bipart,
>     Temp, T;
>     bipart:=true;
>     AB[0] := {};
>     AB[1] := {};
>     R := vertices(G);
>     E := edges(G);
>     v := R[1];
>     AB[0] := AB[0] union {v};
>     i := 0;
>     while R <> {} do
>        T := R intersect AB[i];
>        i := i+1 mod 2;
>        for j from 1 to nops(T) do
>           w := T[j];
>           AB[i] := AB[i]
>             union (neighbors(w, G) minus (AB[0] union AB[1]));
>        od;
>        R := R minus T;
>     od;
>     Temp := SetProd(AB[0]) union SetProd(AB[1]);
>     for i from 1 to nops(E) do
>        if not ({ends(E[i], G)} intersect Temp  = {} ) then
>           bipart := false
>        fi;
>     od;
>     if (bipart = true) then
>        printf('The graph is bipartite with\n');
>        printf('\tbipartition A = %a, B = %a\n',
>                          AB[0], AB[1]);
>     else
>        printf('The graph is not bipartite\n');
>     fi;
>   end:
```

We will now illustrate this procedure for some particular graphs, namely, C_4, C_5, K_4, $K_{5,4}$, and Q_3.

```
>   Bipartite(cycle(4));
```

```
The graph is bipartite with
        bipartition A = {1, 3}, B = {2, 4}
```

```
>   Bipartite(cycle(5));
```

```
The graph is not bipartite
```

```
>   Bipartite(complete(4));
```

```
The graph is not bipartite
```

```
>  Bipartite(complete(4,5));
```

The graph is bipartite with
        bipartition A = {1, 2, 3, 4}, B = {5, 6, 7, 8, 9}

```
>  Bipartite(cube(3));
```

The graph is bipartite with
        bipartition A = {0, 3, 5, 6}, B = {1, 2, 4, 7}

Subgraphs, Unions and Complements

This section focuses on how to use Maple to derive graphs from other graphs. Specifically, we will show how to create subgraphs, graph unions and complements of graphs, using Maple commands and procedures.

We begin with deriving subgraphs from a given graph. To do this, we will induce a subgraph from a given graph by selected a subset of the vertices of a graph, and considering only those edges that have both end-vertices in the selected subset. This can be accomplished by using the induce command, which takes a subset, S of vertices of a graph G, and the graph G, and returns an induced graph H that has a vertex set S and a subset of the edges of G. For illustrative purposes, consider two induced subgraphs on our graph G1, which we will call West and East.

```
>  East:=induce(
>       {Washington,
>        'New York',
>        Detroit,
>        Chicago},
>       G1):
>  West:=induce(
>       {Denver,
>        'San Francisco',
>        'Los Angeles'},
>       G1):
>  draw(East);
```

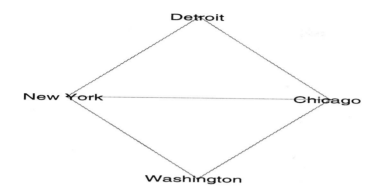

```
>  draw(West);
```

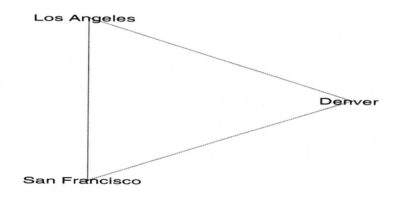

We now examine how to determine if a given graph is a subgraph of another graph. To determine whether a given graph, H, is subgraph of another graph, G, involves nothing more than ensuring that the vertex set of H is a subset of the vertex set of G, and that the edge set of H is a subset of the edge set of G. Specifically, we can create a Maple procedure that will determine subgraph characteristics as follows;

```
>  IsSubgraph:=proc(H::graph, G::graph)
>    local i, is_sub, EG;
>    EG:={};
>    is_sub:=true;
>    for i from 1 to nops(edges(G)) do
>      EG:=EG union {ends(edges(G)[i], G)};
>    od;
>    if not (vertices(H) minus vertices(G) = {}) then
>      is_sub:=false
```

```
>    fi;
>    for i from 1 to nops(edges(H)) do
>      if not (member(ends(edges(H)[i], H), EG)) then
>        is_sub:=false;
>      fi;
>    od;
>    is_sub;
>  end:
```

We now use our newly created procedure on a few examples, including the earlier induced graphs East and West;

```
>  IsSubgraph(East, G1);
```
$$true$$

```
>  IsSubgraph(West, G1);
```
$$true$$

```
>  IsSubgraph(East, West);
```
$$false$$

```
>  IsSubgraph(complete(2), complete(3));
```
$$true$$

```
>  IsSubgraph(cycle(4), complete(4));
```
$$true$$

```
>  IsSubgraph(cycle(4), complete(3));
```
$$false$$

We now move on to consider graph complements. The complement, $\bar{G}$, of a graph G is the graph that has the same vertex set as G, and with the edge set of $\bar{G}$ being the set of all pairs of vertices of G that have no edge between them. In other words, if G has n vertices, then $\bar{G}$ has n vertices and the edge set of $\bar{G}$ is the edge set of K_n with the edge set of G deleted. To construct graph complements using Maple, we simply use the complement function as follows, on the graphs G1 and C_5, as well as on the Petersen graph.

```
>  draw(complement(G1));
```

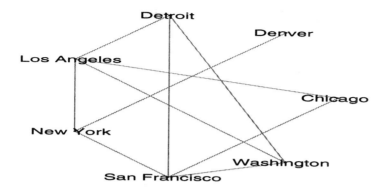

```
>  draw(complement(cycle(5)));
```

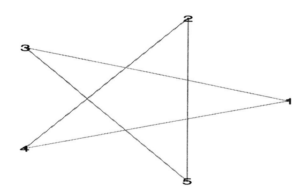

```
>  draw(complement(petersen(10)));
```

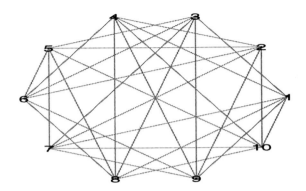

The union of two graphs, as defined on Page 446 of the text, is even easier to use, since Maple has a predefined procedure, called **gunion**, which computes the union of two given graphs. To see the use of this procedure, consider the following;

```
>   G3 := new():
>   addvertex({a,b,c,d,E}, G3);
```
$$a, b, E, d, c$$

```
>   addedge({{a,b},{a,d}, {b, E}, {b,c}, {c,E}, {d,E}}, G3);
```
$$e1, e2, e3, e4, e5, e6$$

```
>   new(G4):
>   addvertex({a,b,c,d,f}, G4);
```
$$a, b, f, d, c$$

```
>   addedge({{a,b}, {b,c}, {b,d}, {b,f},{c,f}}, G4);
```
$$e1, e2, e3, e4, e5$$

```
>   draw(G3);
```

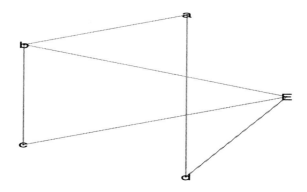

```
>  draw(G4);
```

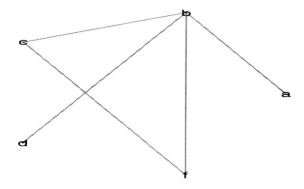

```
>  draw(gunion(G3, G4));
```

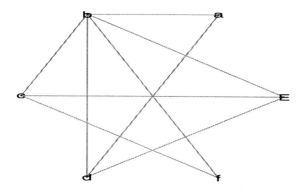

7.4   Representing Graphs, and Graph Isomorphism

Maple allows various techniques and structures to represent alternative forms of graphs. Specifically, in Maple we can represent a graph as a list of vertices adjacent to each other instead of the representation of a set of vertices with a set of interconnecting edges, Maple allows us to construct such a representation easily.

In this section, we will examine how to represent graphs in terms of adjacency lists, adjacency matrices and incidence matrices. We will then use the adjacency matrix representation to help determine whether two graphs are isomorphic.

Representing Graphs

To begin, we will construct a simple graph and determine the adjacency list for the given graph, which consists of lists of those vertices adjacent to each vertex of the graph. Using the Maple **neighbor** function mentioned earlier, we can create a simple procedure to output the adjacency list for a given simple graph. The implementation is as follows.

```
>   AdjList:=proc(G::graph)
>      local i;
>      for i from 1 to nops(vertices(G)) do
>        print('Vertex',
>                vertices(G)[i],
>                'is adjacent to ',
>                neighbors(vertices(G)[i], G));
>      od;
>      end:
>   new(H1):
>   addvertex({a,b,c,d}, H1):
>   addedge({{a,b}, {a,d}, {b,c}, {b,d}, {c,d}}, H1):
>   AdjList(H1);
```

$$Vertex, a, \text{ is adjacent to }, \{b, d\}$$
$$Vertex, b, \text{ is adjacent to }, \{a, d, c\}$$
$$Vertex, d, \text{ is adjacent to }, \{a, b, c\}$$
$$Vertex, c, \text{ is adjacent to }, \{b, d\}$$

For some applications, it is useful to use an adjacency matrix to represent a graph. Once you have defined a particular graph in Maple, you can use the **adjacency** command to find the adjacency matrix of this graph. We will illustrate how to use this command by finding the adjacency matrix of the graph H1. Note that the matrix that is returned has 4 rows and 4 columns, representing the 4 vertices of H1.

```
>  adjacency(H1);
```

$$\begin{bmatrix} 0 & 1 & 0 & 1 \\ 1 & 0 & 1 & 1 \\ 0 & 1 & 0 & 1 \\ 1 & 1 & 1 & 0 \end{bmatrix}$$

We can also find the incidence matrix of a graph previously defined in Maple, using the **incidence** command. We will illustrate the use of this command by constructing the incidence matrix of the graph H1. In this case, notice that the matrix has 4 rows, representing the 4 vertices, and 5 columns, representing the 5 edges of H1.

```
>  incidence(H1);
```

$$\begin{bmatrix} 1 & 1 & 0 & 0 & 0 \\ 1 & 0 & 1 & 0 & 1 \\ 0 & 0 & 1 & 1 & 0 \\ 0 & 1 & 0 & 1 & 1 \end{bmatrix}$$

If we create a more complicated graph, called H2, that has loops and multiple edges but the same vertex set as H1, we can examine how individual entries in the above matrices are changed.

```
>  new(H2):
>  addvertex({a,b,c,d}, H2):
>  addedge( [{a,b}, {a,d}, {b,c}, {b,d}, {c,d},
>            {a,a}, {b,b}, {a,d}, {a,d}, {c,d}],
>  H2);
```

$$e1, e2, e3, e4, e5, e6, e7, e8, e9, e10$$

```
>  incidence(H2);
```

$$\begin{bmatrix} 1 & 0 & 1 & 0 & 0 & 0 & 1 & 0 & 1 & 1 \\ 1 & 0 & 0 & 1 & 1 & 0 & 0 & 1 & 0 & 0 \\ 0 & 1 & 0 & 1 & 0 & 1 & 0 & 0 & 0 & 0 \\ 0 & 1 & 1 & 0 & 1 & 1 & 0 & 0 & 1 & 1 \end{bmatrix}$$

```
> adjacency(H2);
```

$$\begin{bmatrix} 2 & 1 & 0 & 3 \\ 1 & 2 & 1 & 1 \\ 0 & 1 & 0 & 2 \\ 3 & 1 & 2 & 0 \end{bmatrix}$$

Graph Isomorphism

We shall now briefly delve into how to use Maple to examine whether two graphs are isomorphic. The problem of determining whether a graph is isomorphic to another graph is a difficult problem since there is no known set of sufficient conditions for determining the existence of an isomorphism. In other words, there is no set of polynomial-time testable conditions that can ensure that there is an isomorphism between two graphs. It is worth noting that we can exhaustively search through all possible vertex mappings, but this is takes an exponential, not polynomial, amount of time in terms of the size of the graph, since there are $n!$ mappings of vertices in the worst case for a pair of graphs on n vertices. The positive side is that there is a class of necessary conditions that we can use to rule out the existence of isomorphism between two graphs. One set of invariants can be found on Page 455 of the textbook, but this list is rather incomplete in its scope. We shall create a Maple procedure that checks whether a set of invariants of two graphs agree. If they do not then we know that the graphs cannot be isomorphic; if they do, then the graphs may, or may not, be isomorphic. The invariants we use must agree if the two graphs are isomorphic; they use concepts covered throughout Chapter 7 of the text.

```
> Invariants := proc(G::graph, H::graph)
>   local i, not_iso;
>   not_iso:=false;
>   #Ensure that the vertex sets are of the same size
>   if not (nops(vertices(G)) = nops(vertices(H)) ) then
>     not_iso := true;
>   fi;
>   #Ensure that the edge sets are of the same size
>   if not (nops(edges(G)) = nops(edges(H)) ) then
>     not_iso:=true;
>   fi;
>   #Compare degree sequences
>   if not( degreeseq(G) = degreeseq(H) ) then
>     not_iso := true;
>   fi;
>   #Compare edge connectedness
>   if not (connectivity(G) = connectivity(H) ) then
```

```
>       not_iso := true;
>    fi;
>    #Count number of unique spanning trees
>    if not (counttrees(G) = counttrees(H) ) then
>       not_iso := true;
>    fi;
>    #Determine shortest cycle length (girth)
>    if not (girth(G) = girth(H) ) then
>       not_iso := true;
>    fi;
>    if not_iso=false then
>      print('Isomorphism of the graphs is not ruled out');
>      print('by examining this set of invariants.');
>    else
>      print('The graphs are not isomorphic');
>    fi;
>  end:
```

Note that we have made use of a new Maple function, `degreeseq`, which returns an ascending list of degrees of the vertices of the given graph. As examples, consider the earlier defined graphs H1 and H2.

```
>  degreeseq(H1);
```

$$[2, 2, 3, 3]$$

```
>  degreeseq(H2);
```

$$[3, 4, 5, 6]$$

So, using our newly defined procedure, we can rule out existence of isomorphisms for many pairs of graphs. As examples, consider the following.

```
>  Invariants(H1, H2);
```

The graphs are not isomorphic

```
>  Invariants(H1, H1);
```

Isomorphism of the graphs is not ruled out
by examining this set of invariants.

In concluding this section, we remark that the reader can use this procedure along with an exhaustive search technique, to determine whether two graphs are isomorphic. Specifically, if the `Invariants` procedure returns that the two graphs have a possibility of being isomorphic after a series of tests, then we can use the following procedure, called `Isomorph`, to either find an isomorphism or conclude that the two given graphs are not isomorphic. The technique is one of brute force, searching through all possible mappings between the vertex sets of the given graphs. There are somewhat more efficient techniques available, but these are beyond out scope here; consult books on graph algorithms, such as those listed in the Suggested Readings section of the text, for further information.

```
> Isomorph:=proc(G::graph, H::graph)
>   local i, j, n, VH, VG, iso_found, temp, cur_perm, P;
>   #Initialize variables
>   iso_found := false;
>   cur_perm:=1;
>   VH:=[];
>   VG:=[];
>   n:=nops(vertices(G));
>   #Create initial mapping (G[i] -> H[i])
>   for i from 1 to n do
>     VH:=[op(VH), vertices(H)[i]];
>     VG:=[op(VG), vertices(G)[i]];
>   od;
>   #Deduce all permutations (mappings)
>   P:=combinat[permute](VH);
>   while not iso_found do
>     iso_found := true;
>     #Ensure that degrees are same for vertices
>     for i from 1 to n do
>       if not (vdegree(VG[i], G) = vdegree(VH[i],H)) then
>         iso_found:=false;
>       fi;
>     od;
>     #If degrees are same, then compare edges to ensure
>     #that edge in H is an edge in G
>     if iso_found then
>       for i from 1 to n do
>         for j from i to n do
>           if (not (edges({VH[i], VH[j]}, H) = {})) then
>             if edges({VG[i], VG[j]}, G) = {} then
>               iso_found := false;
>             fi;
>           else
>             if (not (edges({VG[i], VG[j]}, G)={})) then
>               iso_found := false;
>             fi;
>           fi;
>         od;
>       od;
>     fi;
>     #Rearrange the vertex mappings
>     cur_perm := cur_perm+1;
>     if cur_perm <= n! then
>       VH:=P[cur_perm];
>     else
>       iso_found := true;
>     fi;
>   od;
>   #Output result
>   if cur_perm > n! then
>     print('There is no isomorphism between the two graphs');
>   else
>     print('An isomorphism is :');
>     for i from 1 to n do
```

```
>           print(VG[i], ' -> ', VH[i]);
>        od;
>     fi;
>   end:
```

In order to test our procedure, we will need two graphs that are isomorphic, but are not trivially isomorphic. To do this, we simply create two new graphs I1 and I2 in the following manner:

```
>   new(I1):
>   new(I2):
>   addvertex({u1, u2, u3, u4, u5}, I1):
>   addvertex({v1, v2, v3, v4, v5}, I2):
>   addedge({ {u1, u2}, {u1, u5}, {u2, u4},
>             {u2, u3}, {u3, u1}, {u3, u4},
>             {u4, u5}}, I1):
>   addedge( { {v1, v2}, {v1, v5}, {v2, v3}, {v2, v5},
>             {v3, v4}, {v4, v1}, {v4, v5}}, I2):
```

We are now ready to test our procedure on these two isomorphic graphs, using our newly created `Isomorph` procedure:

```
>   Isomorph(I1, I2);
```

An isomorphism is :
u4, − , v4
u1, − , v2
u2, − , v5
u5, − , v1
u3, − , v3

7.5   Connectivity

Determining connectivity of graphs can be viewed as a problem of determining whether a city is reachable by certain transportation routes (such as highways) from a given location. Here, we will show how to use Maple to study the connectivity of graphs.

To begin, suppose we are given an undirected graph, and we wish to determine whether or not there is a path between two given vertices of the graph. One method of approaching this problem is to use the Maple `components` command, which returns a set of sets, where each set is a set of vertices in a specific connected components. We may then apply Theorem 1 on Page 464 of the text to determine that there is a path between two vertices if they are in the same connected component. The Maple implementation is illustrated below;

```
> PathExists:=proc(G::graph, v1::name, v2::name)
>   local i, h, H, path_exists;
>   path_exists:=false;
>   H := components(G);
>   for h in H   do
>     if member(v1, h)
>     and member(v2, h) then
>        path_exists:=true;
>     fi;
>   od;
>   path_exists;
> end:
```

We will try this procedure on our graph H1, defined earlier, as well as a new graph, which will have more than 1 component.

```
> new(Uncon1):
> addvertex({a,b,c,d}, Uncon1);
```
$$a, b, c, d$$

```
> addedge({{a,b}, {c,d}}, Uncon1);
```
e1, e2

```
> PathExists(Uncon1, a, c);
```
false

We now extend our notion of connectivity to examine directed graphs. For instance, we are now looking at how to get between two cities using only one-way streets or one-way rail lines. In the directed graph case, we have two levels of connectedness: *strongly connected* graphs involve directed paths between any two vertices, and *weakly connected* graphs involve undirected paths between any two vertices.

To illustrate, we will construct a Maple procedure that determines if a directed graph is strongly connected, weakly connected or not connected at all. The pseudocode for our procedure, which will be called `DiConnect`, is the following.

1. To determine strong connectivity, form spanning subgraphs from each vertex of the original graph. If there is a spanning subgraph that does not contain all vertices of the original graph, then we can deduce that the original graph is not strongly connected. It is worth noting that these spanning subgraphs are spanning trees, which are covered in the next chapter.

2. To determine weak connectivity, we make use of the Maple `gsimp` command, which transforms a directed graph into a simple undirected graph. Having done this, we use the earlier defined function to determine connectivity.

3. If the graph fails to meet any of the criteria of step 1 or step 2, we can deduce that the directed graph is neither weakly nor strongly connected.

In the actual implementation of this algorithm shown below, we are careful to trap any errors that may be generated by attempting to compute them. The Maple implementation of this pseudocode is:

```
>  DiConnect := proc(G::graph)
>    local i, V, strong_con;
>    strong_con := true;
>    V := vertices(G);
>    for i from 1 to nops(V) do
>      if not (traperror(vertices(spantree(G, V[i]))) = V) then
>        strong_con := false;
>      fi;
>    od;
>    if (strong_con = true) then
>      print('The graph is strongly connected');
>    else
>      print('The graph is not strongly connected.');
>    fi;
>    if nops(components(gsimp(G))) = 1 then
>      print('The graph is weakly connected')
>    else
>      print('The graph is not weakly connected')
>    fi;
>  end:
```

Here are three graphs which can be used for testing.

```
>  new(D1): addvertex({a,b,c,d,f}, D1):
>  addedge({[a,b], [a,c], [a,f], [b,d], [c,f]}, D1):
>  new(D2): addvertex({a,b,c,d,f}, D2);
```

$$a, b, f, c, d$$

```
>  addedge( {[a,b], [a,c], [a,f], [b,d],
>    [f,c], [d,a], [c,a]}, D2):
>  new(D3): addvertex({a,b,c,d,f}, D3):
>  addedge({[a,b], [a,c], [a,f], [c,f]}, D3):
>  DiConnect(D1);
```

The graph is strongly connected
The graph is weakly connected

```
>  DiConnect(D2);
```
$$\text{The graph is strongly connected}$$
$$\text{The graph is weakly connected}$$

```
>  DiConnect(D3);
```
$$\text{The graph is not strongly connected.}$$
$$\text{The graph is not weakly connected}$$

Next, we will determine not only if a path exists between two vertices, but how many paths of a given length exist between two vertices in a graph. This problem turns out to be only as complex as matrix multiplication, by Theorem 2 on Page 468 of the text. Simply put, we need to determine the adjacency matrix of a given graph, and then compute the power of that matrix to the exponent equal to the path length. For instance, we can determine that for the undirected graph H1 there are

```
>  A:=adjacency(H1);
```

$$A := \begin{bmatrix} 0 & 1 & 0 & 1 \\ 1 & 0 & 1 & 1 \\ 0 & 1 & 0 & 1 \\ 1 & 1 & 1 & 0 \end{bmatrix}$$

```
>  evalm(A*A*A)[1,3];
```
$$2$$

paths of length 3 between vertex a and vertex c (which are vertices 1 and 3).

In conclusion, we will explicitly determine a path between two vertices, given that one exists. To do this, we will use the Maple **spantree** command for determining a spanning tree of a graph, which we will cover in depth in the next chapter. Briefly, a spanning tree of a graph is a connected graph without cycles that has all the vertices of the original graph. This spanning tree will then be used by the Maple **path** command to find a path between two given vertices. The implementation of our procedure, which will be called **FindPath**, is as follows;

```
>  FindPath:=proc(G::graph, v1::name, v2:: name)
>     local i,c,C,T;
>     #Check to see if a path exists
>     if not PathExists(G, v1, v2) then
>       print('No path exists between the given vertices');
>     else
>       C:=components(G);
>       for c in C do  # check for both vertices
```

```
>               if member(v2, c) then
>                   #Create an induced graph, then determine the
>                   #spanning tree on this induced graph
>                   T:=spantree(induce(c , G), v2);
>               fi;
>           od;
>           print('This is a path between the two vertices');
>           print(path([v1, v2], T));
>       fi;
>   end:
```

```
>   FindPath(Uncon1, a,c);
```
No path exists between the given vertices

```
>   FindPath(Uncon1, a,b);
```
This is a path between the two vertices
[a, b]

7.6   Euler and Hamilton Paths

In this section we will show how to use Maple to solve two problems
that seem closely related, but which are quite different in computational
complexity. The two problems that will be analyzed are the problems of
finding a simple circuit that contains every edge exactly once (an Euler
circuit) and the problem of finding a simple circuit that contains every
vertex exactly once (a Hamilton circuit).

Euler circuits

We will first examine the problem of Euler circuits in a given graph.
Having read through the derivation of Theorem 1 up to Page 476 of
the textbook, we know that a graph has an Euler circuit if, and only
if, every vertex has even degree. Using this result, we can create a simple
Maple procedure for determining whether an undirected graph has an
Euler circuit as follows.

```
>   with(networks):
>   Eulerian := proc(G::graph)
>     local v, V, is_eulerian;
>     if hastype(degreeseq(G),odd) then false else true fi;
>   end:
```

Here, we have made use of existing procedures to produce a list of the
vertex degrees and one of Maple's type testing constructs to search that
list for odd integers.

Having constructed this "helper" procedure, we can implement Algorithm
1 on Page 476 in order to find an Euler circuit, if one exists in the graph.

Note that, in the algorithm, we create a temporary undirected graph H, which is decomposed into Euler circuits, and these circuits are represented by directed edges in the circuit C, which is returned to the user.

```
>  Euler:=proc(G::graph)
>     local i, j, V,C, cur_vertex, sub_start,
>     next_vertex, zero_degree, H;
>     zero_degree:={};
>     V:=vertices(G);
>     if Eulerian(G) = false then
>       print('The graph has no Euler circuit');
>     else
>       new(C);
>       new(H);
>       addvertex(V, C);
>       addvertex(V, H);
>       addedge(ends(edges(G), G),H);
>       #End of initialization
>       cur_vertex:=V[1];
>       #Create first directed Euler circuit in C
>       while not (member(V[1], neighbors(cur_vertex, H))) do
>         next_vertex:=neighbors(cur_vertex, H)[1];
>         addedge([cur_vertex, next_vertex], C);
>         H:=delete(edges({cur_vertex, next_vertex}, H)[1], H);
>         cur_vertex:=next_vertex;
>       od;
>       addedge([cur_vertex, V[1]], C);
>       H:=delete(edges({cur_vertex, V[1]}, H)[1], H);
>       while not (edges(H) = {}) do
>         #While there are still unused edges in H,
>         #create Euler circuits
>         for i from 1 to nops(vertices(H)) do
>           if vdegree(vertices(H)[i], H) = 0 then
>             zero_degree:=zero_degree union {vertices(H)[i]}
>           fi;
>         od;
>       H:=delete(zero_degree, H);
>       #Remove vertices of zero degree from H
>       sub_start:='intersect'(vertices(H), vertices(C))[1];
>       cur_vertex:=sub_start;
>       while not (member(sub_start, neighbors(cur_vertex, H))) do
>         next_vertex:=neighbors(cur_vertex, H)[1];
>         addedge([cur_vertex, next_vertex ], C);
>         #Add the newest edge into C
>         H:=delete( edges( {cur_vertex, next_vertex} ,H)[1], H);
>         #Remove the newest edge from H
>         cur_vertex:=next_vertex;  #Move to the next vertex
>       od;
>       addedge([cur_vertex, sub_start], C);
>       H:=delete(edges({cur_vertex, sub_start}, H)[1], H);
>       od;
>     fi;
>     C;
>  end:
```

We now create a graph and use the algorithm above to form a directed graph which represents the Euler circuits of this graph.

```
>  new(E1):
>  addvertex({1,2,3,4,5}, E1);
```
$$1, 2, 3, 4, 5$$

```
>  addedge({{1,2}, {1,5}, {2,5}, {3,4}, {3,5}, {4,5}},E1);
```
$$e1, e2, e3, e4, e5, e6$$

```
>  ends(edges(Euler(E1)), Euler(E1));
```
$$\{[4,5], [5,3], [5,1], [2,5], [1,2], [3,4]\}$$

If we alter our circuit search to focus on visiting each vertex, rather than each edge, exactly once, the problem generally becomes too difficult to solve computationally, for even relatively small graphs, even in Maple. But, we can implement a brute force method in Maple to determine all possible Hamiltonian paths by constructing all possible permutations of vertices. This implementation of the inefficient algorithm is as follows.

```
>  Ham:=proc(G::graph)
>    local i, ham_path, cur_perm, ham_found, n, P;
>    ham_path:=[];
>    cur_perm:=1;
>    n:=nops(vertices(G));
>    #Create vertex list
>    for i from 1 to n do
>      ham_path:=[op(ham_path), vertices(G)[i]];
>    od;
>    P:=combinat[permute](ham_path);
>    ham_found:=false;
>    while not ham_found do
>      ham_found:=true;
>      for i from 1 to n-1 do
>        if edges({ham_path[i], ham_path[i+1]}, G)={}  then
>          ham_found:=false;
>        fi;
>      od;
>      if not ham_found then
>        #Alter the Hamiltonian path
>        if cur_perm < n! then
>          cur_perm:=cur_perm + 1;
>          ham_path:=P[cur_perm];
>        else
>          ham_found:=true;
>        fi;
>      fi;
>    od;
>    if cur_perm = n! then
>      print('There is no hamiltonian path in the graph');
>    else
```

```
>           print('This is a hamiltonian path in the graph');
>           print(ham_path);
>      fi;
>   end:
```

We will now test this procedure on the complete graph on 4 vertices, along with the graph E1 that we used earlier.

```
>   Ham(complete(4));
```

$$\text{This is a hamiltonian path in the graph}$$
$$[1, 2, 3, 4]$$

```
>   Ham(E1);
```

$$\text{This is a hamiltonian path in the graph}$$
$$[1, 2, 5, 3, 4]$$

7.7   Shortest Path Problems

Among the most common problems in graph theory are the "shortest path problems". Generally, in shortest path problems, we wish to determine a path between two vertices of a weighted graph that is minimum in terms of the total weight of the edges.

This section of this book will focus on representing weighted graphs and using Maple to solve shortest path problems. For instance, we will use the Maple command **shortpathtree** which uses Dijkstra's algorithm, as outlined on Page 493 of the text, to determine a tree containing all shortest paths from a given vertex.

First, we will describe how to represent weighted graphs using Maple. As a first example, we will construct a graph using Maple that is equivalent to the graph shown on Page 492 of the text. The process of graph creation is similar to the unweighted edge case (which is what we have been studying in this chapter up to this point), except for the use of the **addedge** command, which will now make use of optional parameters that have been unused up to this point.

```
>   new(S1):
>   addvertex({a,b,c,d,y,z}, S1);
```

$$c, z, d, a, b, y$$

```
>   addedge(
>      [{a,b}, {a,d},
>       {b,c}, {b,y},
>       {c,z},{d,y},{y,z}],
>      weights=[4,2,3,3,2,3,1],
>   S1);
```

$$e1, e2, e3, e4, e5, e6, e7$$

So, the reader can notice that we simply create the edges in an ordered list, and use the `weights` parameter to assign a weight to each edge created. In order to do something useful with this graph, we will (indirectly) use Dijkstra's algorithm to construct a tree (which is a connected graph that has no cycles) that will contain the minimum distance from a specified vertex to all other vertices.

As an example, if we wish to determine the shortest path from vertex a to vertex z, we can execute the following Maple command, which will construct a tree of the shortest paths from a to all other vertices.

```
>  T1:=shortpathtree(S1, a):
>  draw(T1);
```

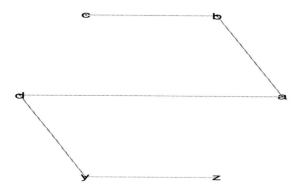

If we specifically wished to extract the shortest path distance from a to z, we can read this directly off the tree, since the weight of each vertex in the returned tree T1 is precisely the shortest distance to vertex a.

```
>  vweight(z, T1);
```

$$6$$

If we wished to not only determine the shortest path from a to all vertices, but rather the shortest distance between every set of vertices, Maple has a simple command, called `allpairs`, that returns a table of elements that contain this information. This specifically implements Floyd's Algorithm as found on Page 498 of the text, and is used as follows;

```
>  allpairs(S1);
```

$$\text{table}([$$
$$(b, y) = 3$$
$$(z, c) = 2$$
$$(b, c) = 3$$

$$(z, a) = 6$$
$$(a, a) = 0$$
$$(c, z) = 2$$
$$(d, z) = 4$$
$$(y, y) = 0$$
$$(b, d) = 6$$
$$(c, y) = 3$$
$$(d, y) = 3$$
$$(a, z) = 6$$
$$(a, y) = 5$$
$$(c, c) = 0$$
$$(y, b) = 3$$
$$(b, z) = 4$$
$$(y, d) = 3$$
$$(y, z) = 1$$
$$(z, b) = 4$$
$$(a, b) = 4$$
$$(z, d) = 4$$
$$(b, b) = 0$$
$$(y, a) = 5$$
$$(d, b) = 6$$
$$(d, d) = 0$$
$$(c, d) = 6$$
$$(b, a) = 4$$
$$(y, c) = 3$$
$$(d, a) = 2$$
$$(z, z) = 0$$
$$(c, a) = 7$$
$$(c, b) = 3$$
$$(d, c) = 6$$
$$(z, y) = 1$$
$$(a, d) = 2$$
$$(a, c) = 7$$
$$])$$

Thus, Maple offers implementations for both Dijkstra's and Floyd's algorithms, as well as the weighted graph constructs that allow any other weighted graph algorithm to be implemented in Maple.

7.8   Planar Graphs and Graph Coloring

This section explains how Maple can be used to determine whether a
graph is planar and will discuss how to use Maple to color a graph. We
begin with planarity of a graph

Planar Graphs

A graph is planar if it can be drawn in the Euclidean plane such that no
two edges meet at a point that is not a vertex. This problem arises in
applications to electronic circuits, where circuit lines are not allowed to
cross due to flow of current needing to be directed properly.

For determining the planarity of graphs, Maple offers the command,
isplanar which returns true if, and only if, the given graph is a planar graph. Its usage is outlined below in a few examples.

```
>  isplanar(complete(4));
```
$$true$$

```
>  isplanar(complete(5));
```
$$false$$

```
>  isplanar(complete(3,4));
```
$$false$$

```
>  isplanar(S1);
```
$$true$$

Graph Colorings

This subsection is focused on the problem of determining how to properly
color a graph: that is, assigning colors to each vertex of a graph such that
no edge will have both end vertices colored by the same color.

Maple is limited by the computational complexity of coloring: it is worth
noting that, in terms of computational complexity, Hamilton circuits and
graph coloring are equivalently difficult problems. Thus, we can either
focus on near optimal colorings, which may use more than the minimum
number of colors, or we can focus our attention on certain classes of
graphs, in the hopes of characterizing colorings in certain cases. In this
section, we will focus our attention on creating a "greedy" algorithm that
colors a graph using no more colors than twice the maximum degree of
the vertices of the graph. (The reader is encouraged to prove this result.)

The pseudocode is as follows.

1. Arbitrarily order the vertices. (Note: Maple does this when the graph is created).

2. Starting at vertex v_1, color it color 1.

3. At step i, color v_i with the smallest unused color amongst its neighbors.

4. Return a pair of results. The first component will contain the total number of colors used, and the second component will contain the actual coloring for the graph.

The implementation of this is as follows;

```
> GreedyColor:=proc(G::graph)
>    local i, j, C, U, V, total_used;
>    V:=vertices(G); total_used:=1;
>    C[V[1]]:=1;
>    for i from 2 to nops(V) do
>      C[V[i]]:=0;
>    od;
>    for i from 2 to nops(V) do
>      #Determine minimum color used to color neighbors of i
>      U:={};
>      for j from 1 to nops(neighbors(V[i], G)) do
>        U:=U union {C[neighbors(V[i], G)[j]]};
>      od;
>      j:=1;
>      while member(j, U) do
>        j:=j+1;
>      od;
>      C[V[i]]:=j;
>      if j>total_used then
>        total_used:=j;
>      fi;
>    od;
>    [total_used, eval(C)];
> end:
```

```
> GreedyColor(complete(4));
```

$$[4, \text{table}([$$
$$1 = 1$$
$$2 = 2$$
$$3 = 3$$
$$4 = 4$$
$$])]$$

```
>  GreedyColor(complete(3,3))[2];
```
$$\text{table}([$$
$$1 = 1$$
$$2 = 1$$
$$5 = 2$$
$$3 = 1$$
$$4 = 2$$
$$6 = 2$$
$$])$$

```
>  GreedyColor(petersen(10))[1];
```
$$3$$

An additional Maple command that is of useful in graph coloring is the command chrompoly that determines the chromatic polynomial of a given graph. This polynomial, when evaluated at a non-negative integer value less than the chromatic number of the given graph, returns a value of zero. This function is used as follows.

```
>  K4:=chrompoly(complete(4), x);
```
$$\mathit{K4} := x\,(x - 1)\,(x - 2)\,(x - 3)$$

```
>  x:=3: K4;
```
$$0$$

```
>  x:=4: K4;
```
$$24$$

7.9   Flows

The final application that we will cover for this chapter concerns a flow on a weighted directed graph. A flow is an assignment of non-negative values to edges such that the value of the arcs entering a vertex is the same as the value of the arcs leaving a vertex, except for two special vertices in the graph. These vertices, called the source and sink, have the property that the only arcs incident with these vertices are outgoing and incoming, respectively. Additionally, the value assigned to a given arc must be less than or equal to the arc weight.

The most common problem when dealing with flows is to determine a maximum flow between two vertices in a weighted directed graph. We will use the Maple procedures called flow and dinic to determine a maximum flow for a given weighted directed graph. In order to use these

commands, we will first create a weighted directed graph, which we will
call **F1**.

```
>  with(networks):
>  new(F1):
>  addvertex({a,b,c,d,x,y,z}, F1):
>  addedge([[a,b], [a,c], [a,x], [b,a], [b,c], [b,z],
             [c,y], [c,d], [d,c], [d,y], [d,x], [x,z],
             [y,z]], weights=[2,4,5,3,1,6,4,8,3,4,2,2,1],
>  F1):
```

We are now ready to determine what the maximum flow is between vertices **a** and **z**, given this weighted graph. One method to do this in Maple
is to use the **flow** command as follows.

```
>  flow(F1, a,z);
```

$$4$$

This result is the actual flow value that is sent by the vertex **a**, and
received by the vertex **z**. If we wish to extract those edges that have a
flow value equal to their edge weight, we can use this **flow** command
again, adding on additional parameters to retrieve this information.

```
>  flow(F1,a,z,cap_edges, cap_vertices);
```

$$3$$

```
>  cap_edges;
```

$$\{\{c, d\}, \{z, x\}, \{y, z\}\}$$

It is worth noting that our parameter `cap_vertices` contains the vertex
set of the edge set `cap_edges`. To see this, we simply type in Maple

```
>  cap_vertices;
```

$$\{a, y, c, d, x\}$$

The **flow** command uses an augmenting path algorithm that is attributed
to Dinic, and this algorithm can be used directly in Maple by the use of
the **dinic** command as follows;

```
>  dinic(F1,a,z,cap_edges2,cap_vertices2);
```

$$3$$

```
>  cap_edges2;
```

$$\{\{c, d\}, \{z, x\}, \{y, z\}\}$$

```
>  cap_vertices2;
```

$$\{a, y, c, d, x\}$$

which returns the same result as the **flow** command, as expected.

7.10 Computations and Explorations

Exercise 1. Display all graphs on four vertices.

Solution: To solve this problem, we use the `powerset` command from the `combinat` package of Maple. Specifically, we shall generate all possible edge sets and then construct the graphs based on these edge sets. The implementation is as follows:

```
>   Prob1:=proc()
>     local i, j, T, edge_set;
>     T:={};
>     #Create all possible edge sets
>     edge_set:=combinat[powerset]({{a,b},{a,c},{a,d},{b,c},{b,d},{c,d}});
>     for i from 1 to 64 do
>       G[i] := new();
>       addvertex({a,b,c,d}, G[i]);
>       addedge(edge_set[i], G[i]);
>       T:=T union {G[i]};
>     od;
>     RETURN(T);
>   end:
```

We now execute this procedure and examine two parts of the output.

```
>   T:=Prob1():
>   draw(T[2]);
```

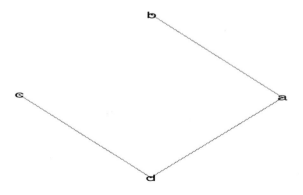

```
>   draw(T[60]);
```

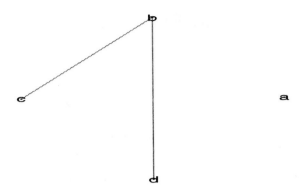

Exercise 2. Display a full set of nonisomorphic simple graphs with six vertices.

Solution: You can solve this problem by first modifying our solution to the previous problem so that it generates all graphs on six vertices. Then, using the **Isomorph** procedure developed in this chapter, you can systematically distinguish the isomorphism classes that this set comprises. (Be prepared to wait a long time for this result.)

Exercise 3. Construct 10 random graphs, and determine their connectivity, maximum degree, minimum degree, greedy coloring number in the average case.

Solution: To solve this question, we first must create a set of random graphs. This can be done by using the **random** command of Maple, which creates a graph on a specified number of vertices with a given edge appearing with a probability of 0.5. This function is used as the following;

```
>  draw(random(4));
```

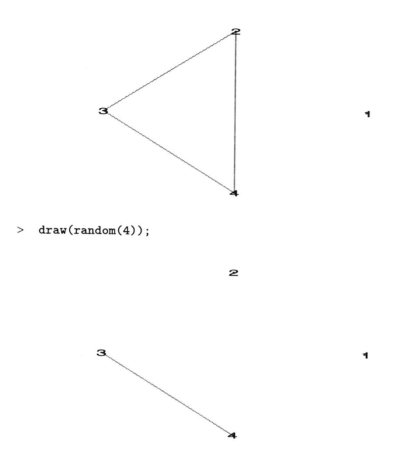

```
>   draw(random(4));
```

To solve the problem at hand, we will form a Maple procedure that will do the following:

1. At step i, where i ranges from 1 to 10, create a random graph
2. Measure the maximum degree using the Maple `maxdegree` command, summing up this in a local variable
3. Measure the minimum degree using the Maple `mindegree`, totaling this result in a local variable
4. Measure the (edge) connectivity of the graph using the Maple command `connectivity`
5. Use the `GreedyColor` algorithm to complete the greedy color, again totally this result
6. Repeat step 1 with i incremented.

An implementation of this pseudocode is as follows:

```
>  RandGraph:=proc(num_vertices::integer)
>     local min_sum, max_sum,
>     color_sum, i, Temp, connect_sum;
>     min_sum:=0;
>     max_sum:=0;
>     color_sum:=0;
>     connect_sum:=0;
>     for i from 1 to 10 do
>       Temp:=random(num_vertices);
>       min_sum:=min_sum+mindegree(Temp);
>       max_sum:=max_sum+maxdegree(Temp);
>       connect_sum:=connect_sum+connectivity(Temp);
>       color_sum:=color_sum+GreedyColor(Temp)[1];
>     od;
>     print('The average minimum degree is',
>          evalf(min_sum/10));
>     print('The average maximum degree is',
>          evalf(max_sum/10));
>     print('The average connectivity is',
>          evalf(connect_sum/10));
>     print('The average greedy chromatic number is',
>          evalf(color_sum/10));
>  end:
>  RandGraph(10);
```

$$\textit{The average minimum degree is}, 2.300000000$$
$$\textit{The average maximum degree is}, 6.600000000$$
$$\textit{The average connectivity is}, 2.300000000$$
$$\textit{The average greedy chromatic number is}, \text{GreedyColor}(\textit{Temp})_1$$

```
>  RandGraph(15);
```

$$\textit{The average minimum degree is}, 3.900000000$$
$$\textit{The average maximum degree is}, 10.40000000$$
$$\textit{The average connectivity is}, 3.900000000$$
$$\textit{The average greedy chromatic number is}, \text{GreedyColor}(\textit{Temp})_1$$

Exercise 4. Determine the average number of planar graphs out of a set of ten random graphs on 20 vertices. Out of this set of graphs, determine the thickness of each non-planar graphs.

Solution: To solve this problem, we use a technique similar to the one use in the previous problem. That is, we will generate ten random graphs on 20 vertices using the Maple random command. On each of these random graphs, we will issue the isplanar command, to determine if a given graph is a planar. The reader will be left to determine the thickness of a non-planar graph, but a general idea to solve this problem is as follows:

1. Given that the graph is non-planar, partition the edges of the graph into i sets.

2. Construct i graphs such that graph G_j has edge set $j, j = 1..i$

3. If one of these graphs is non-planar, then repeat step 2 with a different partition of edges.

4. If there are no more partition of edges, increment i and repeat step 1.

A partial implementation of the first part of this question is as follows.

```
>  Prob3:=proc()
>    local i, Temp, num_planar;
>    num_planar:=0;
>    new(Temp);
>    for i from 1 to 10 do
>      Temp:=random(20);
>      if isplanar(Temp) then
>        num_planar:=num_planar+1;
>      else
>        #Determine thickness of the (nonplanar) graph
>      fi;
>    od;
>    print('The number of planar graphs is', num_planar);
>  end:
```

Exercise 5. Generate random graphs on ten vertices until there is one that has an Euler circuit.

Solution: To solve this problem, we make use of a similar loop structure as in Prob3, except that we are now using the earlier defined `Eulerian` and `Euler` procedures, instead of the `isplanar` function. The solution is the following.

```
>  Prob5:=proc()
>    local i, Temp, euler_found;
>    new(Temp);
>    euler_found:=false;
>    #Loop until the suitable graph is found
>    while not euler_found do
>      Temp:=random(10);
>      if Eulerian(Temp) then
>        euler_found:=true;
>      fi;
>    od;
>    #Extract the Eulerian path
>    Euler(Temp);
>  end:
>  Prob5();
```

$$Euler(\mathit{Temp})$$

Exercise 6. Generate random graphs on ten vertices until there is one that has a Hamilton circuit.

Solution: The solution to this problem is similar to the previous problem, except that we may replace the `Euler` function with the `Ham` function that we created earlier in this chapter.

Exercise 7. Estimate the probability that a randomly generated simple graph with n vertices is connected, for each positive integer n not exceeding ten, by generating a set of random simple graphs and determining whether each is connected.

Solution: This problem involves use of the `random` function, to generate random graphs, as well as the `components` function to determine if a given graph is connected. The implementation is outlined below, with comments to aid in the understanding of the Maple code;

```
>  Prob6:=proc()
>     local i, j, num_connected, Temp;
>     for i from 1 to 10 do
>        #Initialize the counter variable
>        num_connected:=0;
>        for j from 1 to 20 do
>           #Create a random graph the current size
>           Temp:=random(i);
>           #If the graph has one component, it is connected
>           if nops(components(Temp)) = 1 then
>              num_connected:=num_connected+1;
>           fi;
>        od;
>        #Output the result
>        printf('The estimated probability');
>        printf('  that a graph on %d vertices\n',i);
>        printf('is connected is %g\n', num_connected/20);
>     od;
>  end:
>  Prob6();
```

```
The estimated probability  that a graph on 1 vertices
is connected is 1
The estimated probability  that a graph on 2 vertices
is connected is .5
The estimated probability  that a graph on 3 vertices
is connected is .7
The estimated probability  that a graph on 4 vertices
is connected is .6
```

```
The estimated probability  that a graph on 5 vertices
is connected is .75
The estimated probability  that a graph on 6 vertices
is connected is .85
The estimated probability  that a graph on 7 vertices
is connected is .9
The estimated probability  that a graph on 8 vertices
is connected is 1
The estimated probability  that a graph on 9 vertices
is connected is 1
The estimated probability  that a graph on 10 vertices
is connected is 1
```

7.11   Exercises/Projects

Exercise 1. Develop additional procedures for graphically displaying graphs on the screen. Some additional approaches you may want to implement are:

1. drawing the graphs as "planar" as possible by having as few crossing edges as possible

2. drawing the graphs in groups of vertices according to the chromatic number of the graph

3. implementing curved lines to reduce intersecting edges

Exercise 2. Write a Maple program to construct complete n-partite graphs $K_{m_1, m_2, \ldots, m_n}$. That is, graphs where the vertex set is partitioned into m disjoint sets with m_i, $i = 1, 2 \ldots n$ vertices, respectively, and with every vertex in one of subsets of the partition connected to every vertex of the graph not in this subset by an edge.

Exercise 3. Develop a Maple procedure for finding an orientation of a simple graph.

Exercise 4. Develop a Maple procedure for finding the bandwidth of a simple graph.

Exercise 5. Develop a Maple procedure for finding the radius and diameter of a simple graph.

Exercise 6. Make as much progress as possible finding the minimum number of queens controlling an $n \times n$ chessboard for different values of n. See Page 525 of the text for additional assistance.

Exercise 7. Develop a Maple procedure for finding all cliques of a simple graph.

Exercise 8. Develop two different procedures for finding the maximum flow through a weighted graph.

Exercise 9. Use Maple to find as many self-complementary graphs as possible.

Exercise 10. Develop a Maple procedure for graph total colorings, where both edges and vertices are colored.

Exercise 11. Given a sequence of a positive integers, this sequence is called **graphic** if there is a simple graph that has this sequence as its degree sequence. The **degree sequence** of a graph is the nondecreasing sequence made up of the degrees of the vertices of the graph. Develop a Maple procedure for determining whether a sequence of positive integers is graphic and if it is, to construct a graph with this degree sequence.

Exercise 12. Use Maple to construct all regular graphs of degree n, given a positive integer n.

Exercise 13. Use Maple to construct the **odd graph** O_k where k is a positive integer less than 10., The vertices of the odd graph O_k are the subsets of $\{1, 3, \ldots, 2k - 1\}$ and edges connect two vertices if and only if the corresponding subsets are disjoint. Are any of these graphs isomorphic to certain graphs you recognize?

Exercise 14. Write Maple procedures to determine whether an undirected graph has an Euler path and, if so, to find such a path.

Exercise 15. Write Maple procedures to determine whether a directed graph has an Euler circuit and, if so, to find such a circuit.

Exercise 16. Write Maple procedures to determine whether a directed graph has an Euler path and, if so, to find such a path.

Exercise 17. Write a Maple procedure that computes the thickness of a nonplanar graph.

8   Trees

This chapter is devoted to computation aspects of the study of trees. Trees are a specific type of graph, that is connected simple graphs that have no simple circuits.

The Maple code in this chapter assumes that you are using an upgraded version of Maple's networks package. These enhancements primarily affect the display of trees. In particular, the `draw` command has been updated to understand how to draw rooted trees. To test that you are using correct version, load the `networks` package and run the command `version`, as in

```
>  with(networks): version();
```
$$\text{Networks Package Upgrade : Version 4.1()}$$

If this does not produce a description of the version, then you are using the wrong version. An appropriate version can be found at the ftp site at `FTP::/ftp.maplesoft.com/maple/books/rosen` along with installation instructions.

First, we will discuss how to represent, display, and work with trees using Maple. Specifically, we will describe how to represent and construct trees and derive basic characteristics about trees in Maple. We will show how to use Maple to display trees. We will show how to solve a variety of problems, where trees play an important role using Maple, such as in searching and in constructing prefix codes using a specific implementation of the Huffman coding algorithm. We will describe how to use Maple to carry out different methods of traversing trees, where a traversal is a visiting of vertices in some predefined order. Then we will discuss how these traversals relate to the topic of sorting. We continue by showing how to use Maple to create spanning trees of graphs. Then, we will show to use Maple to solve a variety of problems via backtracking. Finally, we will show how to find minimum weight spanning trees of weighted graphs using Maple.

8.1   Introduction to Trees

To begin, we will demonstrate how to construct trees in Maple. Given an unrooted tree, we can construct this tree in Maple just as we would any graph. We will also provide a procedure that uses some built-in capabilities of Maple that determines whether a specific graph is a tree.

Before delving into the implementation, there are two important points that must be stressed. First, we note that Maple differs from the terminology of the text in the sense that Maple refers to simple *cycles* when the text refers to simple *circuits*. The second noteworthy point is that an unrooted tree is a simple graph that has no simple cycles. A rooted tree is exactly the same structurally as an unrooted tree, with the additional property that there is a specific vertex, called the root, which is viewed as the starting point in a tree. In terms of Maple implementation, we represent unrooted trees as graphs, and we create rooted trees from unrooted trees by using Maple commands such as **spantree**, which will be covered later, by specifying a desired root of an unrooted tree.

One other important type of tree is an ordered tree, which is a rooted tree where the children of a vertex of ordered in some manner, such as $1st, 2nd, ..., m-th$ children if there are m children of a given vertex. We will make use of vertex weights to determine the order of children of a specific vertex. This type of tree will arise later, but it is important to distinguish between unrooted trees, rooted and unordered trees, and rooted and ordered trees.

As a first example, we will discuss unrooted trees. We create a tree in exactly the same fashion as we created a graph, using the **networks** package of Maple. As our first example, we will create a simple tree on 4 vertices.

```
>   with(networks):
>   new(T1): addvertex({a,b,c,d,f,g},T1):
>   addedge({{a,b},{a,c},{a,d},{b,f},{b,g}}, T1):
>   draw(Tree(a),T1);
```

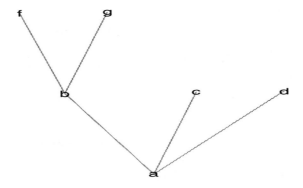

Suppose that we were given a graph and were asked to determine whether or not it was a tree. By the definition of trees, we need to verify the following 3 properties:

1. The graph is connected.
2. The graph is simple.
3. The graph has no cycles.

Using Maple, these properties are easily checked. In particular, we can determine whether a graph is connected in Maple by using the components command, which returns a collection of sets of vertices, where each set in this collection contains the vertices from a connected component of the graph. We can determine whether a graph is simple by using the Maple command gsimp, that returns the underlying simple tree of a multigraph, and then comparing the number of edges of the underlying simple tree to the original graph. This leads to the procedure IsSimple.

```
>    IsSimple := proc(G::graph) local H;
>        H := networks[duplicate](G);
>        if nops(edges(gsimp(H))) = nops(edges(G)) then true
>        else false fi;
>    end:
```

Note that we should not simplify G itself as such a simplification is an irreversible process.

To test for connectivity we provide the procedure IsConnected

```
>    IsConnected := proc(G::graph)
>    evalb(nops(components(G)) = 1) end:
```

We can determine whether a graph has no cycles by using the cyclebase command of Maple that returns a set of cycles, or simple circuits, that form a basis of all cycles (simple circuits) in the given graph; if the cyclebase has no cycles, the graph has no cycles. This, together with the two previous tests can be used to provide a test if a graph is a tree.

```
>    IsTree:=proc(G::graph)
>        if not (IsConnected(G) and IsSimple(G)) then
>          RETURN(false); fi;
>        if cyclebase(G) = {} then RETURN(true);
>        else RETURN(false); fi;
>    end:
```

If you prefer, you can replace the cycle base test in this procedure by one which checks to see if the number of edges is one less than the number of vertices.

We are now ready to use the IsTree procedure to determine whether some particular graphs are trees;

```
>    IsTree(T1);  IsTree(complete(3));
```
                              true

false

Rooted Trees

Up to this point we have dealt with only unrooted trees. We can use the Maple `spantree` command to change an unrooted tree into a rooted tree. It accomplishes this by updating the sets of ancestors and daughters (descendents) for each vertex, to reflect the structure of the spanning tree.

To use the `spantree` command, we select a vertex and form a spanning tree with this vertex as the root, directing all edges in the tree towards the root. (We will study spanning trees later in this chapter. Generally, the `spantree` command takes an undirected connected graph G and a vertex v of the graph and constructs a spanning tree of G using v as the root, directing all edges towards v.) For example, we can make the tree T1 into a rooted tree, taking a as its root using the command

```
>  T2:=spantree(T1, a):
```

We can easily examine relationships between the vertices of a tree using built-in Maple commands. Among the commands that are useful for this are the `daughter`, `ancestor`, `neighbors` and `departures` commands. The `daughter` command finds the children of a vertex in a rooted tree, and the `ancestor` command of Maple finds the parent vertex of a vertex in a rooted tree. The `neighbors` and `departures` act in a similar manner, determining the children of a vertex in the rooted tree.

To illustrate the usage of some of these commands in Maple, we can examine relationships of trees such as parents, children, ancestors and descendants of specific vertices. For instance, we can find the children of the vertex a in the tree T2, using the command:

```
>  daughter(a, T2);
```
$$\{b, c, d\}$$

To find the parent of d in the tree T2, we use the command:

```
>  ancestor(d, T2);
```
$$\{a\}$$

We now present a procedure that finds all the descendants, ancestors, and siblings of a particular vertex in a rooted tree. This procedure, called `Family`, can be described using the following pseudocode:

1. To find all ancestors, we use the `ancestor` command of Maple until there are no more ancestors (i.e. when we reach the root vertex).

2. To find all descendants, we use the `daughter` command repeatedly until there are no more descendants (i.e. when all leaves from a vertex have been reached).

3. To find all siblings of a vertex v, we first find the ancestor of v, called w; the siblings of v are the descendants of w other than v.

An implementation of this procedure is as follows:

```
> Family := proc(v::name,G::graph)
>    local Temp, Ancestors, Descendants, Siblings;
>    Ancestors := ancestor(v,G);
>    Temp := ancestor(v,G);
>    while not (Temp = {}) do
>       Ancestors := Ancestors union Temp;
>       Temp := ancestor(Ancestors,G);
>    od;
>    Descendants := daughter(v,G);
>    Temp := daughter(v,G);
>    while not (Temp = {}) do
>       Descendants := Descendants union Temp;
>       Temp := daughter(Descendants,G);
>    od;
>    Siblings := daughter(ancestor(v, G), G) minus {v};
>    [Ancestors,Siblings,Descendants];
>  end:
```

We will now build a larger tree, called T3 which is the tree shown on **Page 543** of the text, and then we will execute the newly created procedure on one of its vertices.

```
> new(T3):
> addvertex({A,B,C,D,E,F,G,H,I,J,K,L,M,N},T3):
> addedge( {[A,B],[A,J],[A,K],[B,C],[B,E],[B,F],
>    [C,D],[F,G],[F,I],[G,H],[K,L],[L,M],[L,N]}, T3):
> draw(Tree(A),T3);
```

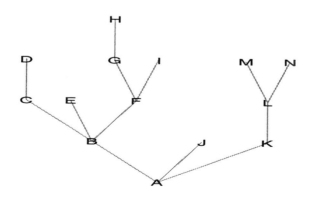

The descendants of the vertex B are obtained by the commands

```
> Bfamily := Family(B,T3);  Bfamily[3];
```

$$Bfamily := [\{A\}, \{K, J\}, \{D, G, H, E, F, i, C\}]$$
$$\{D, G, H, E, F, i, C\}$$

Next, we determine the set of internal vertices and leaves of a rooted tree. Recall that an v is an internal vertex of a rooted tree if v has children, and that v is a leaf vertex of a rooted tree if v has no children. In other words, in any non-trivial rooted tree (i.e. a rooted tree that is more than a single root vertex), the leaves are those with vertex degree 1, and the internal vertices are vertices with vertex degree greater than 1.

Knowing this, we can use the Maple **vdegree** command to determine the set of leaves and the set of internal vertices of a given rooted tree.

```
> Leaves:=proc(T::graph, root::name)
>    select( proc(x,T) evalb( vdegree(x,T) < 2 ) end,
>       vertices(T) minus {root} , T );
> end:
> Internal:=proc(T::graph, root::name)
>    select( proc(x,T) evalb( vdegree(x,T) > 1 ) end,
>       vertices(T) minus {root} , T );
> end:
> Leaves(T2, a); Internal(T2,a);
```
$$\{f, c, g, d\}$$
$$\{b\}$$

We will now discuss how to find the largest number of children of an internal vertex of a rooted tree. Recall that if m is this number, the tree is called an m-ary tree. We will also describe how to determine if an m-ary tree is balanced. Recall that a tree is balanced if all the leaves are at level h or $h - 1$ if a tree has a total of h levels, where the level of a vertex is the length of the unique path from the root to that vertex.

To use Maple for determining whether a tree is an m-ary tree, we can simply look at the degree sequence of the vertices, taking into account that for all vertices except the root, the degree of that vertex is one more than the number of descendants. This can be accomplished by using the **vdegree** command in Maple. To determine whether a tree is balanced, we can use the internal storage structure of a tree in Maple. We will use the fact that Maple stores the level of a vertex in a tree as the vertex weight for that vertex. For instance, if v is a vertex that is at level 3 in a tree, then we can extract this information by using the **vweight** command on the vertex v.

This technique is formalized by the following Maple procedure:

```
> ArityBalanced:=proc(G::graph, Root::name)
>   local Leaf_Depth, V, Max_Children, is_balanced,i;
>   V:=vertices(G); Leaf_Depth:={};
>   is_balanced:=false;
>   for v in V do
>     if (not (v = Root)) and (vdegree(v,G)=1) then
>       Leaf_Depth:=Leaf_Depth union {vweight(v, G)};
>     fi;
>   od;
>   if nops(Leaf_Depth) > 2 then
>     printf('The tree is not balanced\n');
>   elif nops(Leaf_Depth) = 1 then
>     printf('The tree is balanced\n');
>     is_balanced:=true;
>   elif nops(Leaf_Depth) = 2
>   and abs(Leaf_Depth[1] - Leaf_Depth[2]) > 1 then
>     printf('The tree is not balanced\n');
>   else
>     printf('The tree is balanced %a\n', Leaf_Depth );
>     is_balanced:=true;
>   fi;
>   Max_Children:=maxdegree(G)-1;
>   if vdegree(Root, G) > Max_Children then
>     Max_Children:=vdegree(Root, G);
>   fi;
>   printf('The arity of the tree is %d\n', Max_Children);
>   [Max_Children, is_balanced];
> end:


> ArityBalanced(T3, A):

The tree is balanced {}
The arity of the tree is 3
```

We will now use the `ArityBalanced` procedure to verify the formulae on page 541 of the text for full m-ary trees. That is, we will construct a procedure to compute the number of internal vertices and leaves of a given m-ary tree, and compare these quantities as outlined in Theorem 3 and Theorem 4 on page 541 of the text. The procedure called `TheoremVerify` will use

```
> TheoremVerify:=proc(G::graph, Root::name)
>   local internal, m, leaves, n, i, V, is_full_tree;
>   V:=vertices(G);
>   n:=nops(V);
>   i:=0; internal:=0; leaves:=0;
>   is_full_tree:=true;
>   #Use the ArityBalanced procedure to determine arity
>   m:=ArityBalanced(G, Root)[1];
>   while is_full_tree and i<n do
>     i:=i+1;
>     #If there are no children of the vertex, it is a leaf
```

```
>        if nops(daughter(V[i], G)) = 0 then
>            leaves:=leaves+1;
>        #If the number of children is not m, then it is not a
>        #full tree
>        elif not (nops(daughter(V[i],G)) = m) then
>            printf('The tree is not a full tree\n');
>            is_full_tree:=false;
>        #The current vertex is an internal vertex
>        else
>            internal:=internal+1;
>        fi;
>    od;
>    if is_full_tree then
>        printf('Vertices count is %d\n', n);
>        printf('Computed count (m*i+1) is %d\n', m*internal + 1);
>        printf('Leaf count is %d\n', leaves);
>        printf('Computed count ((m-1)*i + 1) is %d\n',
>            (m-1)*internal+1);
>    fi; NULL;
> end:
```

We will use the **TheoremVerify** procedure to verify Theorems 3 and 4 from the text on a full 3-ary tree.

```
> new(Full1):
> addvertex({A,2,3,4,5,6,7,8,9,10}, Full1):
> addedge({{A,2}, {A,3}, {A,4}, {2,5}, {2, 6}, {2,7},
>            {4,8}, {4,9}, {4,10}}, Full1):

> TheoremVerify(Full1, A);
```

```
The tree is balanced
The arity of the tree is 3
Vertices count is 10
Computed count (m*i+1) is 1
Leaf count is 10
Computed count ((m-1)*i + 1) is 1
```

8.2   Application of Trees

This section is concerned with the use of trees in binary search trees. Specifically, we address the use of trees in binary search algorithms as well as the use of trees in Huffman codes. The reason that we wish to use binary trees is that we can use the binary structure of the tree to make binary decisions (i.e. true/false) regarding insertion or search paths.

A tree is called a binary tree if all vertices in the tree have at most two children. In this chapter, we will be using ordered binary trees. The ordering of the vertices is simply a labeling of the children of a vertex as either the left child or the right child, where the left child is regarded as the child that should be visited first, and the right child is the child that should be visited second.

Binary Insertion

A key benefit of ordered binary trees is that the search time required to find a specific element of the tree is logarithmic in the number of vertices of the tree. The major drawback is that the initial insertion of a vertex is much more expensive. We discuss these in greater detail as we go through the actual implementation of a binary insertion algorithm.

We require vertex labels. In Maple we can use the name of the vertex as a label as it can be either an integer or a character string.[1]

A typical vertex in the tree has two descendents (daughters). We must be able to specify which of these two vertices is the left descendent, and which is the right. We can indicate this by using the *weight* of the vertex. In Maple, each vertex has a default weight of 0 as shown by the simple example

```
>  new(g): addvertex(1,2,g):
>  vweight(1,g);
```

$$0$$

We can use the weight of the vertex to specify a left to right ordering. An even simpler solution is simply to agree that the weight of the vertex is its name and to impose an ordering on those names.

To compare two vertex names, we can use a procedure such as

```
>  IsLessThan := proc(a,b) local t;
>     if type( [a,b], [string,string]) then
>         t := sort( [a,b] , lexorder );
>     else
>         t := sort([a,b]);
>     fi;
>     if a = t[1] then true else false fi;
>  end:
```

Using this comparison allows us to generally ignore what type of label is being used.

```
>  IsLessThan(1,2); IsLessThan(b,a); IsLessThan(1,b);
```

$$true$$
$$false$$
$$true$$

---

[1] Vertex names can not begin with the letter "e". This is to avoid possible confusion with the automatically generated edge names, e1, e2, e3 ...

It also makes it easier to change the comparison criteria at some later point without recoding the entire algorithm.

We will also need to be able to find the root of the tree. The following procedure calculates such a root and forces the tree to remember its root so that the computation does not need to be repeated.

```
>  FindRoot := proc(T::GRAPH)
>     local v, V;
>     V := vertices(T);
>     if not assigned( T(Root) ) then
>         for v in V do
>            if indegree(v,T) = 0 then
>               T(Root) := v;   # remember the root
>            fi;
>         od;
>         if not assigned( T(Root) ) then
>            ERROR('no root') fi;
>     fi;
>     T(Root);
>  end:
```

The procedure for constructing an ordered binary tree by insertion is as follows. For simplicity, we use the vertex name as its value when doing comparisons.

1. Given a vertex v to insert into tree T, we need to locate the correct place in the the tree T to insert v

2. If the tree T is empty, insert v as the root

3. Otherwise, make the root of the tree the current vertex cur_vertex and compare v with cur_vertex. If $v =$ cur_vertex you are done.

4. If $v <$ cur_vertex then search the left child, otherwise, search the right child. This is accomplished by changing cur_vertex to be either the left or the right child and comparing the new cur_vertex with v.

5. Eventually, we will not be able to go search in the direction the comparison says we should go. At this point, insert v as the missing child of cur_vertex.

A detailed implementation of the algorithm is as follows.

```
>  Binsertion := proc(T::graph, x::{string,integer})
>     local cur_vertex, V, i, Kids, Left, Right;
>     V := vertices(T);
>     if nops(V) = 0 then
>        addvertex(x, T);
>        T(Root) := x ;   # remember the root for later
>        RETURN( x );
>     fi;
>
>     # We have a rooted tree ...
```

```
>
>        cur_vertex := FindRoot(T);
>        while x <> cur_vertex do
>
>            #  The relative orderings of the descendants and
>            #  x and cur_vertex determine if x can be added
>            #  as a leaf.
>
>            Kids := daughter(cur_vertex,T);
>            Kids := sort( convert(Kids,list) , IsLessThan );
>            Candidates :=
>             sort( [ x, cur_vertex, op(Kids)], IsLessThan );
>
>            # Begin with the easy cases.
>
>            if nops(Candidates) = 2 then
>
>                # no children so just add in new vertex.
>
>                if IsLessThan(x,cur_vertex) then
>                    addvertex(x,weight='Lft',T);
>                else
>                    addvertex(x,weight='Rht',T);
>                fi;
>                addedge( [cur_vertex,x] , T );
>                cur_vertex := x;
>                break;
>
>            elif nops(Candidates)=4 then
>
>                # two descendents so no insertion
>               # at this level ...
>
>                if IsLessThan(x,cur_vertex)
>              then cur_vertex := Kids[1];
>                else cur_vertex := Kids[2]; fi;
>                next;
>
>            elif nops(Candidates) = 3 then
>
>                #  not this level if pattern is
>                #  [x,L,cur_vertex] or [L,x,cur_vertex]
>                #  [cur_vertex,L,x] or [cur_vertex,x,L]
>
>                if Candidates[1] = cur_vertex
>                or Candidates[3] = cur_vertex
>                then
>                    cur_vertex := Kids[1];
>                    next;
>                fi;
>
>                #  For all remaining cases
>                #  add in x as a new vertex
>
```

```
>                    if IsLessThan(x,cur_vertex) then
>                         addvertex(x,weight='Lft',T);
>                    else
>                         addvertex(x,weight='Rht',T);
>                    fi;
>
>               #  Yes! This level.
>
>                    addedge( [cur_vertex,x] , T);
>                    cur_vertex := x;
>                    break;
>
>               fi;
>          od;
>          RETURN( cur_vertex );
>    end:
```

The `addvertex` procedure is used here in a manner which updates the weights of each vertex as it is created. This is to indicate if the vertex is a left or right descendent. While we do not use these weights, they could be used as an alternative measure for sorting the descendents of any particular vertex.

Instead, for sorting we use the names of the vertices to indicate the relative ordering of vertices. The fact that any two vertices (not just descendants of the same vertex) can be compared allows us to combine the new vertex, the descendents, and the current vertex all into one sorted list which is then inspected to determine which of the many special cases is relevant during the insertion of a new vertex. Whatever method of comparing vertices is used, it is important that the comparison procedure be passed on to the `sort` routine as an extra argument.

To validate this procedure, examine how the following list of integers is added:

```
>   Num_List:=[4,6,2,8,5,3,7,1]:
>   new(Tree_Num):
>   for i from 1 to 8 do
>     Binsertion(Tree_Num, Num_List[i]);
>   od;
```

$$
\begin{array}{c}
4 \\
6 \\
2 \\
8 \\
5 \\
3 \\
7 \\
1
\end{array}
$$

To see the resulting tree and its structure, use the Maple `draw` command.

```
>  draw(Tree(4), Tree_Num);
```

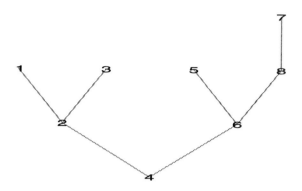

The result is clearly a binary search tree.

Binary trees exist to be searched. The following procedure `BiSearch` does so. Print statements have been added to illustrate the path that the algorithm uses while searching the tree.

```
>  BiSearch := proc(T::graph, v)
>      local  i, Kids, cur_vertex;
>      cur_vertex := FindRoot(T);
>      while v <> cur_vertex do
>          print(cur_vertex);
>
>          # check the easy cases
>
>          if v = cur_vertex then RETURN(true); fi;
>          Kids := daughter(cur_vertex,T);
>          if Kids = {} then RETURN( false) fi;
>
>          # descendants so start looking ...
>
>          Kids := sort( convert(Kids,list) );
>          Candidates :=
>            sort( [v , cur_vertex, op(Kids)], IsLessThan);
>
>          if nops(Candidates) = 4 then # both descendents
>
>            if IsLessThan(cur_vertex,v) then
>                cur_vertex := Kids[2];
>            else
>                cur_vertex := Kids[1];
>            fi;
>              next;   # back to top of loop
>
```

```
>          elif nops(Candidates) = 3 then
>
>              # not present unless cur_vertex
>              # is the first or the last in the list
>
>              if Candidates[1] <> cur_vertex
>              and Candidates[3] <> cur_vertex then
>                  RETURN( false );
>              fi;
>              cur_vertex := Kids[1];
>              next;
>          fi;
>      od;
>      RETURN(true);
>  end:
```

To test this procedure, we try searching for two elements, one of which is in the tree and one of which is not. **Tree_Num**;

```
>  BiSearch(Tree_Num,8);
```
$$4$$
$$6$$
true

```
>  BiSearch(Tree_Num,12);
```
$$4$$
$$6$$
$$8$$
false

Huffman Coding

Huffman coding is a method for constructing an efficient prefix code for a set of characters. It is based on a greedy algorithm, where at each step the vertices with the least weight are examined. Huffman coding can be shown to produce optimal prefix codes. The following pseudocode describes an algorithm for Huffman coding. (For a fuller discussion of Huffman coding, see Cormen, Leiserson, and Rivest, *Introduction to Algorithms*, MIT Press, 1989.)

1. Begin by creating an ordered list of the elements to be coded, where the ordering is with respect to the frequency of occurrence of these elements. Consider each element of the list as a vertex with weight equal to its frequency of occurrence.

2. Remove the first two elements, x and y, from this list

3. Assign x as the left child and y as the right child of a new vertex z in our tree

4. Assign the weight of z to be the sum of the weights of x and y
5. Insert z into the correct position of our list and repeat step (2).
6. Upon completion, our list contains only one element, which is a rooted binary tree.

Once again, a special comparison routine is required. Code elements are represented by lists such as $[a, 15]$ and $[b, 10]$. The following procedure HuffCompare compares two such elements.

```
>  HuffCompare :=proc(a::list,b::list)
>  if a[2] <= b[2] then true else false fi;
>  end:
```

For example, we find that $[b, 10] < [a, 15]$.

```
>  HuffCompare([b,10],[a,15]);
```
$$true$$

Using this method of comparison, lists of code elements can be sorted into ascending order.

```
>  sort( [[a,5],[b,10],[c,8],[d,11]], HuffCompare);
```
$$[[a, 5], [c, 8], [b, 10], [d, 11]]$$

The full Huffman Encoding algorithm is implemented as follows:

```
>  Huffman:=proc(L::listlist)
>     local i, j, k, n, Q, T, x, y, z, Temp;
>     new(T);
>     Q := sort( L , HuffCompare ); i := 1;
>     while(nops(Q)>1) do
>         i := i+1;
>
>         # get the first two code elements
>
>         x:=Q[1]; Q:=subsop(1=NULL, Q);
>         y:=Q[1]; Q:=subsop(1=NULL, Q);
>
>         # build the new vertex and its location
>
>         z := [ i , x[2]+y[2]];
>         for j to nops(Q) while HuffCompare( z, Q[j]) do
>             j := j+1;
>         od;  j := j-1;
>
>         # add the vertices and edges to the tree
>
>         Q := [seq(Q[k],k=1..j),z,seq(Q[k],k=j+1..nops(Q))];
>         addvertex([x[1],y[1],z[1]],weights=[x[2],y[2],z[2]],T);
>         addedge({[z[1],x[1]],[z[1],y[1]]},T);
>     od;
>     RETURN( eval(T) );
```

```
>  end:
```

The type `listlist` denotes a list of lists. The final **eval** is included to ensure that the result returned by the procedure is the actual tree and not just its name.

Try this newly created procedure on the following list of English characters paired with a relative frequency of occurrence;

```
>  Huf:=Huffman([[f,15],[b,9],[d,22],[c,13],[a,16],[e,45]]):
```

To see the result we again use the Maple **draw** command;

```
>  rt := FindRoot(Huf);
```
$$rt := 6$$

```
>  draw(Tree(rt), Huf);
```

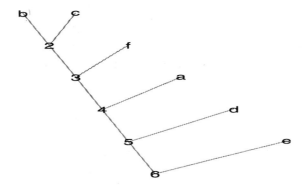

8.3   Tree Traversal

In this section we show how to use Maple to carry out tree traversals. Recall that a tree traversal algorithm is a procedure for systematically visiting every vertex of an ordered rooted tree. In particular, we will provide procedures for three important tree traversal algorithms: preorder traversal, inorder traversal, and postorder traversal. We will then show how to use these traversal methods to produce the prefix, infix, and postfix notations for arithmetic expressions.

These tree traversals rely on the construction of ordered rooted trees. We shall use vertex weights to represent the order of children, as was done in earlier sections.

We create an ordered tree, based on the tree of Figure 3 on page 556 of the text, in order to see the behavior of the various tree traversals. Then, the graph becomes d := 'd':

```
>  new(Trav):
>  addvertex( [a,b,c,d,e,f,g,h,i,j,k,l,m,n,o,p],
>    weights=[0,1,2,3,1,2,1,2,3,1,2,1,2,1,2,3], Trav):
>  addedge( [[a,b],[a,c],[a,d],[b,e],[b,f],[d,g],[d,h],
>    [d,i],[e,j],[e,k],[g,l],[g,m],[k,n],[k,o],[k,p]], Trav):
>  draw(Tree(a),Trav);
```

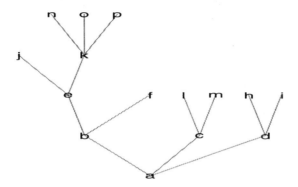

The weights added to the vertices here represent the number of that vertex in terms of its parent. For example, d is the 3rd child of a. As in the previous section, such weights could be used for sorting, though in fact the underlying ordering is simply alphabetic on the names of the vertices.

We first implement the the *preorder traversal algorithm*. This visits the root and then each subtree left to right in a preorder traversal manner. In pseudocode form, the algorithm is;

1. Visit the root of T. Set current vertex to be the root.
2. Consider the children of the current vertex as roots of trees $T_1, T_2, ...T_m$, taken in left to right order. Repeat step (1) on each of these subtrees in order.

As can be seen from step (2) of the above pseudocode, this algorithm is recursive. We provide the following implementation in Maple:

```
>  Preorder:=proc(G::graph, r)
>    local Dep, v, T;
>    #Visit the root
>    printf('%a ', r);
>    #Consider children of the root
>    Dep:= departures(r, G);
```

```
>      #Form the correct order of children
>      Dep:= sort( convert(Dep,list) , IsLessThan );
>      #Preorder traverse these subtrees in order
>      for v in Dep do
>         Preorder(G, v);
>      od;
>      printf('\n', r);
>   end:
```

The order in which we traverse the descendents is determined by the boolean procedure `IsLessThan` (from in the previous section). To traverse the descendants in a different order, use a different comparison routine.

We can examine the execution of this procedure on the earlier created tree `Trav`, rooted at vertex A.

```
>   Preorder(Trav, a);
```

```
a b e j
k n
o
p
f
c l
m
d c l
m
h
i
```

We implement *inorder traversal*, in a similar manner. We simply alter the sequence in which the vertices are visited. Specifically, we look at leftmost subtree vertices, followed by the root, followed by the rightmost subtree vertices. In pseudocode this is:

1. If the tree T has only one vertex, r then visit r
2. Else, the tree T has more than one vertex. Call the leftmost subtree (rooted at the leftmost child) T_1. Inorder traverse T_1, then visit the root r of T.
3. Inorder traverse the rooted subtrees $T_2, ..., T_m$.

A Maple implementation is as follows:

```
>   Inorder:=proc(G::graph, r)
>      local v, Dep, T;
>      # If we have reached a leaf vertex, print it
>      if outdegree(r, G) = 0 then
>         print(r);
>      # We are at an internal vertex
>      else
>         Dep:=departures(r, G);
>         # Determine order of children and traverse the subtree
```

```
>        # based at the leftmost child.
>        Dep := sort( convert( Dep , list ), IsLessThan );
>        Inorder(G, Dep[1]);
>        # Visit the root
>        print(r);
>        # Inorder traverse the remaining subtrees
>        for v in Dep[2..nops(Dep)] do
>           Inorder(G, v);
>        od;
>     fi; NULL;
>  end:
```

We have added a NULL as the last statement so that nothing is returned.

Once again, to traverse the children in a different order, use a different comparison routine to sort the descendants. Test this newly created procedure by executing it on the tree **Trav**.

```
>  Inorder(Trav, a);
```

$$
\begin{array}{c}
j \\
e \\
n \\
k \\
o \\
p \\
b \\
f \\
a \\
l \\
c \\
m \\
l \\
c \\
m \\
d \\
h \\
i
\end{array}
$$

The final traversal that we shall implement in the *postorder traversal*. Postorder traversal is similar to the preorder traversal, except that we visit the root after we visit each subtree. In pseudocode this is:

1. Consider the children of the root as subtrees $T_1, T_2, ...T_m$, taken in left to right order.
2. Postorder traverse T_1, then T_2, up to T_m.
3. Print the root of the current tree.

A Maple implementation of this procedure is as follows:

```
>  Postorder:=proc(G::graph, r::name)
>     local v,i, Dep, T;
>     #Consider children of the root
>     Dep:=departures(r, G);
>     #Form the correct order of children
>     Dep:= sort( convert(Dep,list) , IsLessThan) ;
>     #Postorder traverse these subtrees in order
>     for v in Dep do
>        Postorder(G, v);
>     od;
>     #Visit the root
>     printf(' %c\n', r);
>  end:
```

We also test this procedure on the tree **Trav** with root **A**.

```
>  Postorder(Trav, a);
```

```
j
n
o
p
k
e
f
b
l
m
c
l
m
c
h
i
d
a
```

Infix, Prefix and Postfix Notation

We will now discuss how to use Maple to work with the infix, prefix, and postfix forms of arithmetic expressions. These forms are discussed in Section 8.3 of the text. Specifically, we will show how to create a binary tree representation of an infix expression, how to use postorder and preorder traversals to create postfix and prefix forms of the expression, respectively, and how to evaluate these expressions from their postfix and prefix forms.

To begin this section, we construct a Maple procedure that takes a infix arithmetic expression and converts it into a binary tree representation. This binary tree representation can then be traversed using the traversals of the previous sections to form various arithmetic representation formats. As an example, we can construct a binary tree representation of an infix expression, such as $(3+10)^2 - (100-30)/(5*2)$, then execute a preorder traversal to form a prefix representation of this arithmetic expression.

In the Maple procedure that we will implement, we will consider a modified version of infix notation, to avoid complicated string manipulation that detract from the underlying problem.

Specifically, we will use lists braces [] to indicate normal parenthesis () of infix notation.

Additionally, we will define our operators in terms their string representations: "+" for addition, "-" for subtraction and so on.

Maple does not yet provide good facilities for representing and manipulating strings. However, there are several helpful techniques that we can use to make the notation less intrusive. We represent expressions such as $x + y$ as $[x, +, y]$ where "+" is a string.

Maple procedures can be used to help us to construct such lists. For example,

```
>   '&Plus' := proc(a,b) [a,Unique('+'),b] end:
```

allows us to write and use

```
>   x &Plus y;
```

$$[x, +, y]$$

Maple handles procedure names of the form &... as infix operators. The procedureUnique has is defined especially for the routines used in this book and is part of the the support library for this book that is available via ftp.

The result is a list including a specially encoded [2] version of the string "+" which does not collide with prior uses of "+" as a name. (Every vertex of the expression tree must have a different name, even if they look the same.)

Similarly we use

```
>   '&Times' := proc(a,b)  [a,Unique('*'),b] end:
>   '&Pow'   := proc(a,b)  [a,Unique('^'),b] end:
>   '&Div'   := proc(a,b)  [a,Unique('/'),b] end:
>   '&Minus' := proc(a,b)  [a,Unique('-'),b] end:
```

---

[2]The special procedure Unique ensures that this use of '+' as a name is distinct from prior uses as a name. For example, Unique(a)+Unique(a) does not simplify because the two resulting "a"s are in fact different to Maple.

so that we can write and use

```
> x &Times y, x &Pow y;
```

$$[x, *, y], [x, \hat{\ }, y]$$

These can be used to write the arithmetic expression $((x+y)^2)+((z-4)/3)$ as

```
> Expr1:= (((x &Plus y) &Pow 2) &Plus ((z &Minus 4) &Div 3 ));
```

$$Expr1 := [[[x, +, y], \hat{\ }, 2], +, [[z, -, 4], /, 3]]$$

The result is a nested list of lists, each list representing a binary operation.

Now we are ready to construct a binary tree representing such an expression. The required algorithm in pseudocode is:

1. If there is a single algebraic expression (such as a name or a number) the tree consists of a single vertex.

2. Otherwise, the list consists of a left operand, an operator and a right operand. Use the algorithm to build a binary tree for the left operand

3. Repeat step (2) on the right operand.

4. Combine the results of steps (2) and (3) to form the binary tree.

Our implementation is as follows:

```
> InFixToTree:=proc(L::{list,algebraic})
>    local r1,r3, T1, T3,LocL;
>    # If we have no sublists in our expression, return
>    # the vertices and edge lists as shown
>    if type(L,algebraic) then
>       new(T1); LocL := Unique(L);
>       addvertex(LocL,T1);
>       T1(Root) := LocL;
>       RETURN( eval(T1) );
>    fi;
>
>    # L is now a list such as [a , + , b]
>
>    T1 := InFixToTree(L[1]); r1 := T1(Root);
>    T3 := InFixToTree(L[3]); r3 := T3(Root);
>
>    # construct the new tree
>
>    addvertex(vertices(T3),T1);
>    addedge( ends(T3), T1 );
>    addvertex( L[2] , T1 ):
>    addedge( { [L[2],r1], [L[2],r3] } , T1 );
>    T1(Root) := L[2];
>    RETURN( eval(T1) );
> end:
```

The root of the tree is handled in a special way. Recomputing the root of the tree is awkward and expensive so we simply ask each tree to remember its own root. This is accomplished by assignment statements such as `T(Root) := LocL;`.

The procedure `Unique` is used once again to ensure each instance vertex name is unique, even if it looks the same as another.

In all other aspects, the implementation is almost exactly as outlined in the pseudocode.

To test this procedure, use it to build an expression tree for the earlier expressionExpr1.

> `Expr1;`

$$[[[x, +, y], \hat{\ }, 2], +, [[z, -, 4], /, 3]]$$

> `TreeExpr1:=InFixToTree(Expr1):`

To see this, draw it as a tree.

> `r := TreeExpr1(Root);`

$$r := +$$

> `draw(Tree(r),TreeExpr1);`

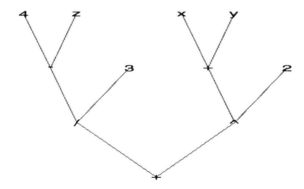

Suppose we are given a binary tree representation of an arithmetic expression. We can use the earlier created algorithms to express these trees as either postfix or prefix expressions by executing the postorder or preorder traversals, respectively. Since this is straightforward, it is left up to the reader to explore this technique.

As a final example in this section, we demonstrate how to evaluate a given postfix expression. For simplicity, we have represented the the basic operations by the first letters of the words add, subtract, multiply, divide, and exponentiate. The reader is left to explore how to implement a procedure that will evaluate a prefix expression, since the technique is a simple modification of the argument that will be used in the postfix case.

```
>  Post1:=[7,2,3,M,S,4,E,9,3,D,A];
```
$$Post1 := [7, 2, 3, M, S, 4, E, 9, 3, D, A]$$

```
>  Postfix:=proc(T::list)
>    local i, L;
>    L:=T;
>    while nops(L)>1 do
>      i:=1;
>      while not member(L[i], {'A','S','D','M','E'}) do
>        i:=i+1;
>      od;
>      if L[i]='A' then
>        L[i]:= L[i-2]+L[i-1];
>      elif L[i]='M' then
>        L[i]:= L[i-2]*L[i-1];
>      elif L[i]='S' then
>        L[i]:= L[i-2]-L[i-1];
>      elif L[i]='D' then
>        L[i]:= L[i-2]/L[i-1];
>      elif L[i]='E' then
>        L[i]:= L[i-2]^L[i-1];
>      fi;
>      L := [op(L[1..i-3]),op(L[i..nops(L)])];
>    od;
>    L;
>  end:
>  Postfix(Post1);
```
$$[4]$$

Note that in release 4, we are permitted to assign directly to list elements.[3]

8.4   Trees and Sorting

This section explains how to use Maple to carry out and analyze sorting algorithms. Trees are often used to model sorting algorithms, especially when the complexity of these algorithms is being studied. In particular, this section focuses on two of the many different sorting algorithms will be studied, the bubble sort and merge sort.

---

[3]In Release 3, you must use `subsop(i=...,L)` in place of `L[i] := ...` , and `[L[i..n]]` in place of `L[i..n]`.

Bubble Sort

To begin, we will examine an implementation of bubble sort. The reason why bubble sort has been given the title of "bubble" is that the smallest of the list "bubble" towards the front of the list, moving one step closer to their position after each iteration. The pseudocode, which is outlined on page 575 of the text, is as follows;

1. Receive as input a list, L, of n elements
2. Loop on index i from 1 to $n - 1$
3. Loop on index j from 1 to $n - i$
4. If the element at position $j + 1$ in the list L is smaller than the element at position j of L, swap these two elements

At the end of each j loop, we placed the largest i elements at the end of L.

The following procedure, called `BubbleSort`, is an implementation of this pseudocode.

```
>   BubbleSort:=proc(L::list)
>     local i, j, temp, T;
>     T:= array(L);
>     for i from 1 to nops(L)-1 do
>       for j from 1 to nops(L)-i do
>         if T[j] > T[j+1] then
>           temp:=T[j];
>           T[j]  := T[j+1];
>           T[j+1] := temp;
>         fi;
>       od;
>     od;
>     convert(T,list);
>   end:
```

Note that before starting to move elements around, we converted the list L to an array. This is because we can change each element of an array in one operation whereas to change any part of a list, we must recopy the entire list – a process involving n operations. When we are finished moving elements around we turn the array back into a list.

We can examine the execution of this procedure on an unsorted list. Note that if the list has five elements, a total of $5 * 4/2 = 10$ loop steps are used by the bubble sort algorithm to order these elements.

```
>   BubbleSort([3,2,4,1,5]);
```
$$[1, 2, 3, 4, 5]$$

Merge Sort

We now will implement the merge sort in Maple. We will also use Maple to study the complexity of this algorithm. The merge sort algorithm can be implemented as a recursive procedure. The rough outline of the task of merge sorting a list is to split the list into two list of equal, or almost equal, size, sorting each sublist using the merge sort algorithm, and then merge the resulting lists. This can be described in pseudocode as follows:

1. Given a list of elements, if the list length is 1, return this list
2. If the list has more than two elements, Merge sort these two lists and return the merged list of the two (as the pseudocode on page 577 of the text outlines).

First we provide a procedure, Merge, that takes two sorted lists and merges them into a single sorted list, consisting of the elements in the two lists. Here is the Maple code for the Merge procedure:

```
>  Merge := proc(L1::list, L2::list)
>    local  L, i,j,k,m,n;
>    L:=[];
>    i := 1; j := 1;  k := 1;
>    m := nops(L1); n := nops(L2);
>    L := array(1..m+n);
>    while  i <= m and j <= n   do
>        if L1[i] <= L2[j] then
>                L[k] := L1[i];
>                i := i+1;
>        else
>                L[k] := L2[j];
>                j := j+1;
>        fi;
>        k := k+1;
>    od;
>    while i <= m do
>        L[k] := L1[i];
>        i := i+1; k := k+1;
>    od;
>    while j <= n do
>        L[k] := L2[j];
>        j := j+1; k := k+1;
>    od;
>    convert(L,list);
>  end:
```

We illustrate the use of this procedure with the following example:

```
>  Merge([1,2,6,8],[3,5,7]);
```

$$[1, 2, 3, 5, 6, 7, 8]$$

We now provide the pseudocode for the merge sort algorithm.

The description of the merge sort algorithm that we will use is based on a recursive definition. In pseudocode,

1. If the list, L has only one element, it is sorted in order, so we return L as is.

2. If L has more than one element, we split the lists into two lists of the same size or where the second list has exactly one more element than the first.

3. We recursively merge sort these two lists, and merge the two sorted lists together.

```
>  MergeSort:=proc(L::list)
>    local First, Second,i, n;
>    # If the list has only one element, return it
>    if nops(L) = 1 then
>      #print(L);
>      L;
>    else
>      # The list has more than one element
>      #print(L);
>      n := nops(L);
>      mid := floor(n/2);
>      # Split the lists into two sublists of equal size
>      First  := L[1..mid];
>      Second := L[mid+1..n];
>      # Merge the result of the Merge sorts of the two lists
>      Merge(MergeSort(First), MergeSort(Second));
>    fi;
>  end:
```

We illustrate the use of the `MergeSort` procedure by sorting an unsorted list with 10 elements:

```
>  MergeSort([8,2,4,6,9,7,10,1,5,3]);
```

$$[1, 2, 3, 4, 5, 6, 7, 8, 9, 10]$$

We will now analyze the running time for `MergeSort` in relation to the running time for `BubbleSort`. Specifically, we will create a 1000 element unsorted list with random elements, and execute both `BubbleSort` and `MergeSort` on this list. This will provide a limited illustration of the running time of these procedure, which the reader should expand upon by reading the theoretical analysis of the text.

To create a random list of 100 elements, we use the `rand` and `seq` commands of Maple as follows:

```
>   A:=[seq(rand(), i=1..100)]:
```

Then, we can use the `time` command to measure the amount of time required to sort the random list:

```
>   st:=time(): BubbleSort(A): time() - st;
```
$$1.167$$

```
>   st:=time(): MergeSort(A): time() - st;
```
$$.600$$

The reader is encouraged to implement other sorting algorithms using Maple and to study the relative complexity of these algorithms when used to sort lists of various lengths. It is also interesting to use animation techniques to illustrate the steps of sorting algorithms. Although we do not do that here, the reader with advanced programming skills is invited to do so. Take special note of the importance of using the right data structure, i.e., lists versus arrays.

8.5   Spanning Trees

This section explains how to use Maple to construct spanning trees for graphs and how to use spanning trees to solve many different types of problems. Spanning trees have already been used in Chapter 7; they have a myriad of applications. In particular, we will show how to use Maple to form spanning trees using two algorithms: depth-first search and breadth-first search. Then we will show how to use Maple to do backtracking, a technique based on depth-first search, to solve a variety of problems.

To begin, we will discuss how to implement depth-first search in Maple. As the name of the algorithm suggests, vertices are visited in order of increasing the depth of the spanning tree. The pseudocode is:

1. Given a graph G, and a root vertex v, consider the first neighbor of v, called w. Add edge (v, w) to the spanning tree.

2. Pick x to be a neighbor of w that is not in the tree. Add edge (x, w) and set w equal to x.

3. Repeat step (2) until there are no more vertices not in the tree.

The depth first search implementation is as follows:

```
>   Depth := proc(G::graph, r)
>      local v, V, N, S,In_Tree;
>      new(S);
>      addvertex(r, S);
```

```
>      In_Tree:=[r];
>      while In_Tree <>[] do
>        v := In_Tree[-1];
>        N:=neighbors(v, G) minus vertices(S);
>        if N = {} then In_Tree := In_Tree[1..nops(In_Tree)-1];
>          next; fi;
>        addvertex(N[1],S);
>        addedge([v,N[1]],S);
>        In_Tree:=[op(In_Tree), N[1]];
>      od;
>      eval(S);
>      end:
```

We demonstrate the usage of the depth-first search procedure with the following example:

```
>   new(G1):
>   addvertex({A,B,C,D,E,F,G,H,I,J,K,L,M}, G1):
>   addedge( {{A,B},{A,D},{B,C},{B,E},{C,F},{D,E},
>      {D,H},{E,F},{E,I},{F,G},{F,J},{G,L},
>      {G,J},{H,K},{H,I},{I,J},{I,K},{M, K}}, G1);
```

$e1, e2, e3, e4, e5, e6, e7, e8, e9, e10, e11, e12, e13, e14, e15, e16, e17, e18$

```
>   S1:=Depth(G1,E):
>   draw(Tree(E), S1);
```

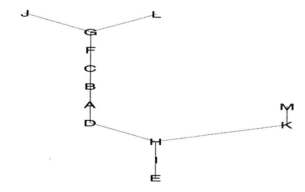

Having implemented the depth-first spanning tree algorithm, we can now modify the Maple code slightly and get a breadth-first spanning tree. Specifically, the breadth-first search algorithm operates by examining all vertices at the current depth of the graph before moving downwards to the next level of the graph. Before implementing this algorithm, we give a pseudocode description of the algorithm.

1. Given a graph G, and a root vertex v, identify the neighbors of v. Call this neighbor set N_1.

2. Add edges from v to each vertex in N_1 that is not already in the spanning tree.

3. Pick the first vertex from N_1, called w. Consider the neighbors of w; call this set of neighbors N_2.

4. Repeat step (2) with w substituted in for v, and N_2 substituted in for N_1.

5. If all vertices in N_1 have been exhausted, move down to the next level, and repeat step (2).

A Maple implementation is the following, called `Breadth`;

```
>   Breadth:=proc(G::graph, r)
>     local v, N, S, In_Tree;
>     new(S);
>     addvertex(r, S);
>     In_Tree:=[r];
>     while not(In_Tree=[]) do
>       v := In_Tree[1];
>       N:=neighbors(In_Tree[1], G) minus vertices(S);
>       for v in N do
>         addvertex(v,S);
>         addedge([In_Tree[1], v],S);
>         In_Tree:=[op(In_Tree), v];
>       od;
>       In_Tree:= In_Tree[2..nops(In_Tree)];
>     od;
>     eval(S);
>     end:
>   S2:=Breadth(G1, E):
>   draw(Tree(E), S2);
```

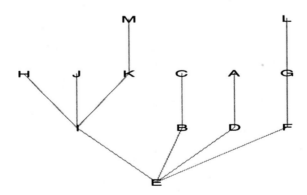

Notice that the two spanning trees are different even though they are rooted at the same vertex. In particular, the depth-first search tree has a deep and thin structure, whereas as the breadth-first search tree has a shorter and wider structure. These graphical representations help to illustrate the algorithm used, and heuristically, we can use the representations to guess whether a depth-first search or a breadth-first search has been used.

Backtracking

Backtracking is a method that can be used to find solutions to problems that might be impractical to solve using exhaustive search techniques. Backtracking is based on the systematic search for a solution to a problem using a decision tree. (See the text for a complete discussion.) Here we show how to use backtracking to solve several different problems, including coloring a graph, solving the n-queens problem of placing n queens on a $n \times n$ chessboard so that no queen can attack another queen, and solving the subset sum problem of finding a subset of a set of integers whose sum is a given integer.

The first problem we will attack via a backtracking procedure is the problem of coloring a graph using n colors, where n is a positive integer. Given a graph, we will attempt to color it using n colors in a greedy manner, as done in the section on Graph Coloring in the text. However, when we reach a coloring that does not allow us to color an additional vertex properly, we backtrack, changing the color on an earlier colored vertex and trying again. Here is the pseudocode for our `BackColor` procedure which carries out this coloring based on backtracking. Here we order the colors as color 1, color 2,..., color n.

1. Order the vertices of the graph G as $v_1, v_2, ..., v_m$.
2. Assign color 1 to v_1. Set $i = 2$
3. Assign color c to v_i, where c is the smallest integer so that no neighbor of v_i has been assigned color c
4. If we can assign such a color to v_i, increment i and repeat step (3).
5. If we cannot assign any color to v_i, we backtrack, setting $i = i - 1$ and incrementing the color of v_i if possible.
6. If we do not have a valid coloring, repeat step (5).
7. Stop when we color all vertices, or we have exhausted all possible colorings.

An implementation of this pseudocode in the following Maple algorithm called `BackColor`:

```
>   BackColor := proc(G::graph,n::integer)
>   local i,k, v, V, cur_vertex, Assigned, Available,
>           used , N, cur_color;
```

```
>    V:= convert(vertices(G), list );
>    #Initialize the Assigned and Available colors
>    for v in V do
>      Assigned(v):=0;
>      Available(v):=[seq(k, k=1..n)];
>    od;
>    cur_vertex:=1;
>    while cur_vertex >= 1 and cur_vertex <=nops(V) do
>      v := V[cur_vertex];
>      # Assign smallest color to current vertex
>      # Gather all neighbors of current vertex
>      N:=neighbors(v, G);
>      while Assigned(v)=0 and Available(v) <> [] do
>        Used := map( Assigned , N );
>        if not member( Available(v)[1], Used ) then
>            Assigned(v) := Available(v)[1];
>        fi;
>        Available(v) := Available(v)[2..nops(Available(v))];
>      od;
>      # Backtrack if no such color exists
>      if Assigned(v) = 0 and (Available(v) = []) then
>        printf('Backtracking on %a %d\n',
>            v, Assigned(v));
>        while (Available(v)= []) and cur_vertex > 1 do
>            Available(v) := [seq(k, k=1..n)];
>            Assigned(v) := 0;
>            cur_vertex := cur_vertex - 1;
>            v := V[cur_vertex];
>        od;
>        if cur_vertex > 1 then
>            Assigned(v) := 0;
>        else
>            break;
>        fi;
>      else
>        cur_vertex:=cur_vertex+1;
>      fi;
>    od;
>    if not has( map( Assigned , V ), 0 ) then
>      for v in V do
>        printf('Assign vertex %a color %d\n', v, Assigned(v));
>      od;
>    else
>      printf('There does not exist a proper vertex coloring\n');
>      printf('with %a colors\n', n);
>    fi;
>  end:
```

We will now try this implementation on a new graph called **C1**. Notice that the output of the BackColor procedure is the current assignment of colors at any backtracking stage, and the final coloring or indication of the non-existence of proper coloring upon termination of the procedure.

```
>  new(C1): addvertex([E,B,C,D,A], C1):
>  addedge({{A,B},{A,E},{B,C},{B,D},{B,E},{C,D},{D,E}},C1):
>  BackColor(C1,3);

Assign vertex D color 1
Assign vertex A color 1
Assign vertex E color 2
Assign vertex B color 3
Assign vertex C color 2
```

Next, we will examine the execution of the BackColor procedure on C1 with two new edges added. Notice that there this new graph has K_4 as a subgraph;

```
>  addedge({{A,D},{A,C}}, C1):
>  BackColor(C1,3);

Backtracking on B 0
Backtracking on B 0
Backtracking on E 0
There does not exist a proper vertex coloring
with 3 colors

>  BackColor(C1,4);

Assign vertex D color 1
Assign vertex A color 2
Assign vertex E color 3
Assign vertex B color 4
Assign vertex C color 3
```

Another problem with an elegant backtracking solution is the problem of placing n-queens on an $n \times n$ chessboard so that no queen can attack another. This means that no two queens can be placed in the same horizontal, vertical, or diagonal line. We will solve this problem using a procedure based on backtracking. We will place queens on the chessboard in a greedy fashion, until either all the queens are placed or there is no available position for a queen to be placed without sitting on the same diagonal, vertical or horizontal line, with a queen that has already been placed.

To make the main procedure easier to understand, we will create a helper procedure that will verify whether a particular placement of queens is valid. If there are two queens on the same row, column or diagonal, then ValidQueens will return false; otherwise, the procedure will return true;

```
>  ValidQueens:=proc(Q::matrix,
>   row::integer, col::integer, size::integer)
>   local i,return_value;
>    return_value:=true;
```

```
>    # Verify the dimensions are valid
>    if row > size or col > size then
>      return_value := false;
>    else
>      # Check Queens horizontally
>      # Note that main algorithm never places two queens
>      # in the same column, so vertical check is not needed
>      for i from 1 to col-1 do
>         if Q[row, i] = 1 then
>            return_value:=false;
>         fi;
>      od;
>      # Check Queens on the two diagonals
>      for i from 1 to col-1 do
>        if row>i then
>           if Q[row-i, col-i] = 1 then
>              return_value:=false;
>           fi;
>        fi;
>        if row+i <=size then
>           if Q[row+i, col-i] = 1 then
>              return_value:= false;
>           fi;
>        fi;
>      od;
>    fi;
>    # Return the value
>    return_value;
>  end:
```

The main procedure for solving the n-queens problem, which will be called
NQueens, follows the same control flow as the BackColor procedure, as
can be deduced from the in-line comments. Specifically, we have an initial-
ization stage, an incremental stage where we try to fill the current column,
and a backtracking stage where we backtrack if we cannot place a queen in
the current column. The Maple implementation of this procedure follows:

```
>  NQueens:=proc(n::integer)
>    local cur_col, cur_row, Q, bad_position, Assigned;
>    # Initialize Queens
>    Q:=linalg[matrix](n, n, 0);
>    cur_col:=1;  Assigned:=[];
>    while cur_col >= 1 and cur_col <=n do
>      # Assign a Queen to the next column
>      bad_position := true;
>      cur_row:=0;
>      # does first available position work?
>      while cur_row < n and bad_position do
>        cur_row := cur_row+1;
>        bad_position := false;
>        # bad if there is a neighbor vertex colored
>        Q[cur_row, cur_col] := 1;
>        if not ValidQueens(Q, cur_row, cur_col, n) then
```

```
>          bad_position := true;
>          Q[cur_row, cur_col] := 0;
>        fi;
>      od;
>      # Backtrack if no available Queen position
>      if cur_row=n and bad_position then
>        printf('Backtracking on column');
>        printf(' %d of %a since stuck\n', cur_col, Q);
>        while not ValidQueens(Q, cur_row, cur_col, n)
>        and cur_col > 1 do
>          cur_col := cur_col-1;
>          Q[Assigned[cur_col], cur_col]:=0;
>          cur_row := Assigned[cur_col] + 1;
>          Assigned:=subsop(cur_col=NULL, Assigned);
>        od;
>        if cur_col >= 1 and cur_row <= n then
>          Assigned:=[op(Assigned), cur_row];
>          Q[cur_row, cur_col] := 1;
>          cur_col := cur_col + 1;
>        else
>          cur_col := cur_col - 1;
>        fi;
>      else
>        # If Queen placement is currently valid,
>        # move to the next column
>        cur_col:=cur_col+1;
>        Assigned:=[op(Assigned), cur_row];
>      fi;
>    od;
>    if (cur_col >= 1) then
>      printf('A proper Queen placement is %a\n', Q);
>    else
>      printf('No Queen placement with %d Queens\n', n);
>    fi;
>  end:
```

We now use the NQueens procedure to solve the n-queens problem when $n = 3$ and $n = 4$;

```
>  NQueens(3);

Backtracking on column 3 of Q since stuck
Backtracking on column 2 of Q since stuck
Backtracking on column 3 of Q since stuck
No Queen placement with 3 Queens

>  NQueens(4);

Backtracking on column 3 of Q since stuck
Backtracking on column 4 of Q since stuck
A proper Queen placement is Q
```

We consider a third problem which can be solved using backtracking; the subset sum problem. Given a set of integers S, we wish to find a subset

B of S such that the sum of the elements of B is a given value M. To use
backtracking to solve this problem, we successively select integers from S
until the sum of these elements equals M or exceeds M. If it exceeds M,
we backtrack by removing the last element in the sum, and we insert a
different value.

Before we implement the main procedure, we will create two small helper
functions that aid in the manipulation of lists. The first helper procedure,
which is called `ListSum` determines the sum of the elements in a given
list.

```
> ListSum:=proc(S::list, Ind::list) local i, T;
>    T:=0;
>    for i from 1 to nops(Ind) do
>      T:=T+S[Ind[i]];
>    od;
>    T;
> end:
```

The second helper function, called `ListInd` determines a subset of a given
list S that is indicated by the positions stored in list J.

```
> ListInd:=proc(S::list, J::list) local i, T;
>    T:=[seq(S[J[i]],i=1..nops(J))];
> end:
```

The main procedure to determine a possible solution to the subset sum
problem, called `SubSum`, follows;

```
> SubSum:=proc(S::list, M::integer)
>    local CurSub, next_index, T, Ind, CurSum,i;
>    # Initialize variables
>    Ind:=[];
>    CurSum:=0;
>    i:=1;
>    next_index:=0;
>    T:=S;
>    # Loop until we reach the given sum value
>    while not (CurSum = M) do
>      printf('The current subset %a has sum %d\n',
>             ListInd(T, Ind), CurSum);
>      next_index:=next_index+1;
>      # If we have reached an impasse, backtrack
>      if next_index > nops(T)
>      and Ind[nops(Ind)] = nops(T) then
>        Ind:=subsop(nops(Ind)=NULL,Ind);
>        Ind:=subsop(nops(Ind)=NULL,Ind);
>        CurSum:=ListSum(T, Ind);
>      else
>        # if out of values to sum, backtrack
>        if next_index > nops(T)  then
>          next_index:=Ind[nops(Ind)]+1;
>          Ind:=subsop(nops(Ind)=NULL, Ind);
```

```
>         CurSum:=ListSum(T,Ind);
>       fi;
>       # If the current subset less than M, then
>       # we add the next value to the subset
>       if CurSum+T[next_index] < M then
>         Ind:=[op(Ind), next_index ];
>         CurSum:=ListSum(T, Ind);
>       fi;
>     fi;
>     # If we have exhausted the index, set variables to
>     # halting values
>     if Ind=[] then
>       T:=subsop(1=NULL, T);
>       break;
>     fi;
>   od;
>   # Return the list sum
>   ListInd(T,Ind);
> end:
```

We execute this procedure on the Example 6 on page 588 of the text:

```
> SubSum([31,27,15,11,7,5], 39);
```

```
The current subset [] has sum 0
The current subset [31] has sum 31
The current subset [31] has sum 31
The current subset [31] has sum 31
The current subset [31] has sum 31
The current subset [31, 7] has sum 38
The current subset [31, 7] has sum 38
The current subset [31, 5] has sum 36
```

⬚

The three problems we have attacked using backtracking, coloring graphs, the n-queens problem, and the subset sum problem are representative of the vast number of problems that can be solved using backtracking and the reader will certainly find occasions when the techniques of this section will help solve such problems. (See Exercise 7 at the end of this section, for example.)

8.6   Minimum Spanning Trees

This section explains how to use Maple to find the minimum spanning tree of a weighted graph. Recall that a minimum spanning tree T of a weighted graph G is a spanning tree of G with the minimum weight of all spanning trees of G. The two best known algorithms for constructing minimum spanning trees are called Prim's and Kruskal's algorithms (although they have an older history); we will develop Maple procedures that implement both of these algorithms here.

We will begin by studying Prim's algorithm, whose pseudocode is outlined on page 594 of the text. Prim's algorithm proceeds by constructing a tree by successively selecting a minimum weight edge that extends this tree from its current set of vertices. The pseudocode is as follows:

1. Start to build the minimum weight spanning tree T with an edge of minimum weight of the entire graph.

2. Add to T the edge of minimum weight that is incident to a vertex in T which does not form a simple circuit in T

3. Repeat step (2) until we have a total of $n - 1$ edges in T

To simplify our implementation of Prim's algorithm, we first create a procedure, MinWeight, that determines the edge of minimum weight with exactly one vertex in a given set of vertices.

```
>  MinWeight:=proc(G::graph, S::set)
>    local e, i, Candidates, Del, Min_Edge;
>    #Determine the set of adjacent edges
>    if S=vertices(G) then Candidates:=edges(G)
>    else Candidates := incident(S,G); fi;
>    if Candidates = {} then RETURN(NULL) fi;
>    # Determine the minimum weight edge candidate
>    Min_Edge:=Candidates[1];
>    for e in Candidates do
>      if eweight(Min_Edge,G) > eweight(e ,G) then
>        Min_Edge:=e;
>      fi;
>    od;
>    RETURN(Min_Edge);
>  end:
```

The special case of all vertices of G is included to provide a convenient starting point Prim's algorithm. In this case, we simply return the edge of G of minimum weight. In all other cases, the search is restricted to edges emanating from the subgraph induced by the specified vertices. The implementation depends on the fact that the procedure incident finds all the edges that leave a particular set of vertices. Also, the overall efficiency of the algorithm can be improved by systematically working our way through a sorted list of edges rather than searching anew for edge candidates at every step.

Given the procedure MinWeight, the actual implementation of Prim's algorithm is straightforward. we first initialize the minimum weight tree T to be the tree with just one edge, an edge of least weight. At each step we add an edge of minimum weight that incident with the current tree T.

```
>   Prim := proc(G::graph)
>     local i, VT, V, T, e;
>     new(T);
>     V := vertices(G);
>     # Add minimum weighted first edge
>     e := MinWeight(G,V);
>     addvertex(ends(e, G), T);
>     addedge(ends(e,G), T);
>     # Loop until all n-1 edges are added to tree
>     for i from 2 to nops(V)-1 do
>       e := MinWeight(G,vertices(T));
>       if e = NULL then ERROR('no spanning tree') fi;
>       # Add new vertex as well as new edge
>       addvertex(ends(e,G) minus vertices(T), T);
>       addedge(ends(e,G),weights=eweight(e,G),T);
>     od;
>     RETURN( eval(T) );
>   end:
```

We return `eval(T)` rather than just `T` to ensure that the actual tree, rather than just its name is passed back.

To test the procedure `Prim` we find a minimum weight spanning tree of the weighted graph from Example 1 on page 595 of the text. You can construct the graph using the commands

```
>   new(City1):
>   addvertex({sf,chic,den,ny,atl},City1):
>   addedge( [{sf,ny},{sf,chic},{sf,den},{sf, atl}],
>     weights=[2000,1200,900,2200], City1):
>   addedge( [{den,chic},{den,ny},{den, atl}],
>     weights=[1300,1600,1400], City1):
>   addedge( [{chic,ny},{chic,atl}, {atl, ny}],
>     weights=[1000,700, 800], City1):
```

Then the minimum weight spanning tree is T1 given by

```
>   T1 := Prim(City1):
```

This tree is best viewed as a tree by selecting a particular root and then drawing the tree.

```
>   draw(Tree(sf), spantree(T1,sf));
```

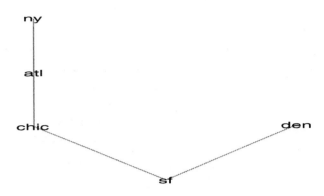

The total weight of its edges can be computed as

```
>   total := 0:
>   for e in edges(T1) do
>     total := total + eweight(e,T1)
>   od:
>   total;
```

2901

Kruskal's algorithm algorithm builds the minimum weight tree by successively adding an edge of least weight that does not form a simple circuit in any of the previously constructed tree fragments. The psuedocode for this algorithm is:

1. Sort the edges of the graph in ascending order.
2. Choose smallest weight edge, e
3. If e creates a cycle in T when added, discard e from the list and repeat step (2)
4. Add e to the minimum weight spanning tree T
5. Repeat step (2) until the tree has $n - 1$ edges

Before we can implement Kruskal's algorithm, we need to be able to sort edges. As in earlier sections, we can do this using Maple's built in sorting routine by providing a suitable procedure for comparison of any two edges.

The comparison routine required here is subtly more complicated than before because it must use the graph in addition to the edge names inside the comparison. This can be accomplished using a template procedure as follows. A specific graph is substituted for a placeholder in a template.

```
>   edgecompare := proc(G::graph)
>     subs({TESTG=eval(G)} ,
>        proc(a,b)
>             if eweight(a,TESTG) <= eweight(b,TESTG) then
>             true else false fi;
>        end );
>   end:
```

By invoking this procedure on a specific graph such as `City1`, we create a comparison procedure customized to that graph.

```
>   comp1 := edgecompare(City1):
```

It can be used as

```
>   comp1(e1,e2);
```

$$false$$

Now to sort a list of the edges of `City1` by weight all we need do is

```
>   edgelist := convert(edges(City1),list);
```
$$edgelist := [e1, e2, e3, e4, e5, e6, e7, e8, e9, e10]$$

```
>   edgelist := sort(edgelist,comp1);
```
$$edgelist := [e9, e10, e3, e8, e2, e5, e7, e6, e1, e4]$$

The weights of this sorted list are in ascending order, as verified by mapping the `eweight` command onto the list.

```
>   map( eweight , edgelist , City1);
```
$$[700, 800, 900, 1000, 1200, 1300, 1400, 1600, 2000, 2200]$$

```
>
```

Armed with this sorting routine, we are nearly ready to implement Kruskal's algorithm. At each step of the algorithm, we have an edge e, a collection of trees T, formed from edges of G, and G, and we must determine if the edge e forms a cycle. This is done by finding the components of T, and checking each component to see if both ends of the edge e are in that same component. This is accomplished by the procedure

```
>   InComponent := proc(e,T::graph,G::graph)
>      local c,C;
>      C := components(T);
>      for c in C do
>        if ends(e,G) minus c = {} then
>             RETURN(true); fi;
>      od;
>      RETURN(false);
>   end:
```

It makes use of the fact that the `components` commands represents each component by a set of vertices.

Now we are ready to implement Kruskal's algorithm.

```
>  Kruskal:=proc(G::graph)
>    local E,T,i,n,e;
>
>    E := convert( edges(G), list); # sort the edges
>    E := sort( E, edgecompare(G));
>
>    #  start building the forest
>    new(T); i := 0; n := nops(vertices(G)):
>    while i < n  and E <> [] do
>      e := E[1];
>      if InComponent( e , T , G ) then
>          E := subs(e=NULL,E);
>          next;
>      fi;
>
>      #  add new edge to forest
>      addvertex(ends(e,G),T);
>      addedge(ends(e,G),T);
>      i := i+1;
>      E := subs(e=NULL,E);
>    od;
>    eval(T);  # the new tree
>  end:
```

This algorithm can also be tested on the tree City1. from Example 1 on page 595.

```
>  T2 := Kruskal(City1):
>  draw(Tree(sf), spantree(T2,sf));
```

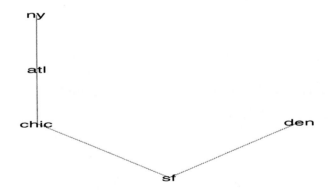

8.7   Computations and Explorations

Exercise 1. Display all trees with six vertices

Solution: To solve this problem, we use a recursive definition of trees. We know that an empty graph is a tree, and that a graph with a single vertex is a tree. We can then build up larger trees from these smaller trees by taking each vertex and forming a new tree by adding a leaf connected to that vertex. (The reader can verify that this truly creates all trees with one more vertex).

Thus, we shall create a Maple procedure, called **ExtendTree**, that takes a set of trees on n vertices, and adds a new edge to each tree, returning the resulting set of trees on $n + 1$ vertices. The maple implementation is as follows:

```
>   ExtendTree:=proc(Trees::set)
>     local i, j, S, t, num_vertices, X;
>     S:={};
>     # Loop over all trees in given set
>     for i to nops(Trees) do
>       T := Trees[i];
>       # Add new vertex
>       num_vertices:=nops(vertices(T));
>       addvertex(num_vertices+1, T);
>       # For each vertex, add new leaf edge
>       for v in vertices(T) do
>         new(X[i][v]);
>         X[i][v]:=duplicate(T);
>         addedge([v , num_vertices+1], X[i][v]);
>         S:=S union {X[i][v]};
>       od;
>     od;
>     S;
>   end:
```

We will now illustrate how to form all trees on 4 vertices, and leave the determination of all trees of larger size to be determined by the reader;

```
>   new(StartingTree):
>   addvertex(1, StartingTree):
>   X:=ExtendTree(ExtendTree(ExtendTree({StartingTree}))):
>   draw(Tree(1),X[1]);
```

```
>  draw(Tree(1), X[2]):
>  draw(Tree(1),X[3]):
>  draw(Tree(1), X[4]):
>  draw(Tree(1),X[5]):
>  draw(Tree(1), X[6]):
```

Exercise 2. Construct a Huffman code for the letters of the English language based on the frequence of their occurrence in ordinary English text

Solution: This problem can be broken down into two smaller problems. The first problem is to determine how to gather the frequency of occurrence for each letter of the English language. The second problem is how to construct a Huffman code based on this frequency of occurrence. We already have created the **Huffman** procedure in Maple that can be used to determine the correct Huffman code given the frequency of occurrence of the English characters. Hence, we have solved the second problem.

To solve the first problem, we can use Maple to scan through a string of text and count the number of occurrences of each letter of the English alphabet. Specifically, we can use strings in Maple in the following manner. Suppose that we have a passage of text which is in lower case and has no punctuation, such as:

```
>  input_text:=
>    'the quick brown fox sat down and had lunch with me';
```

input_text := the quick brown fox sat down and had lunch with me

Then, we can initialize a table indexed on each character of the English language, and then scan through the **input_text** and count the occurrence of each character.

```
>  # Initialization
```

```
> alphabet:='a bcdefghijklmnopqrstuvwxyz';
```
$$alphabet := a\ bcdefghijklmnopqrstuvwxyz$$

```
> for i from 1 to length(alphabet) do
>    freq[substring(alphabet, i..i)]:=0;
> od:
> # Count occurrence of each character
> for i from 1 to length(input_text) do
>    freq[substring(input_text, i..i)] :=
>         freq[substring(input_text, i..i)] + 1;
> od:
> freq[a];
```
$$3$$

```
> freq[e];
```
$$2$$

```
> freq[q];
```
$$1$$

To determine the frequency of occurrence of English letters in certain contexts we can run this program on large sample input. We can simply extend our alphabet to include punctuation and any other special characters that are used in the character set. You will find somewhat different frequency distribution for different types of content, such as literature, correspondence, computer programs, electronic mail, and so on. It is worth noting that many books on cryptography (such as *Cryptography: Theory and Practice* by Douglas R. Stinson, CRC Press, 1995) contain the frequencies of characters in English, and of many other languages. Additionally, this code can be used to count frequency of occurrence of any character set, such as the ASCII character set, the French character set, the Spanish character set, and so on.

Exercise 3. Compute the number of different spanning trees of K_n for $n = 1, 2, 3, 4, 5, 6$. Conjecture a formula for the number of such spanning trees whenever n is a positive integer

Solution: This problem can be solved quite easily using the `counttrees` command of Maple, which returns the number of unique spanning trees of an undirected graph. Thus, to determine the number of unique spanning trees on $K_n, n = 1..6$ we can execute the following Maple statements:

```
> counttrees(complete(1));
```
$$1$$

```
>   counttrees(complete(2));
                                    1
>   counttrees(complete(3));
                                    3
>   counttrees(complete(4));
                                    16
>   counttrees(complete(5));
                                   125
>   counttrees(complete(6));
                                  1296
```

We leave it to the reader to conjecture a formula. A useful hint is to look for a formula of the form $n^{f(n)}$, where $f(n)$ is a simple function in terms of n.

Exercise 4. Compute the number of different ways n queens can be arranged on an $n \times n$ chessboard so that no two queens can attack each other for all positive integers n not exceeding 10.

Solution: This problem can be solved by altering the procedure **NQueens** that was implemented in this chapter. Specifically, when a solution is determined, rather than exiting the procedure, we simply backtrack on that solution, and continue, until all solution paths have been examined. Thus, the procedure will exit only when all solutions have been outputted. We leave it to the reader to alter the **NQueens** procedure and conjecture a formula for the number of solutions in terms of n for the n-queens problem.

Exercise 5. Draw the complete game tree for a game of checkers on a 4×4 board.

Solution:
We will offer a partial solution to this problem; the reader is left to complete the full solution. Specifically, we will create a Maple procedure called **MovePiece** that will determine all possible new checker arrangements given a specific piece that is to be moved on a given checker arrangement. Once this procedure is created, the reader must determine how to represent these board positions as vertices and edges, how to determine the next level of the game tree, as well as any necessary halting conditions.

The implementation of MovePiece is straightforward: we examine each piece that can be moved, and then determine if we can move the piece forward and to the left or right, depending on the piece's current board position and whether there is a piece occupying a possible new position. Also, we will determine if a piece can jump and capture an opponents piece, depending on the board space and position of the opponent's positions. Additionally, we will examine whether a piece is a king, in which case, the piece can move both forwards and backwards on the checkerboard.

We now give the Maple implementation of MovePiece. In-line comments are provided to make this code easier to understand.

```
>   MovePiece:=proc(A::matrix, piece::integer)
>     local i, j, k, cur_column, is_king, S, Temp, direction;
>     # Initialize values, depending on the piece value
>     S:=[];
>     if piece = 1 then direction:=-1;
>     else direction:=1; fi;
>     # Examine all possible positions on board
>     for i from 1 to 4 do
>       for j from 1 to 4 do
>         # If we have found a piece, determine whether
>         # or not it is a King
>         if abs(A[i,j])=piece then
>           if A[i,j] < 0 then is_king:=1;
>           else is_king:=0; fi;
>           # If the piece is a King, then examine both forward
>           # and reverse directions
>           for k from 0 to is_king do
>             if k>0 then direction:=-1*direction; fi;
>             # Examine possible new positions to see if they
>             # are still on the board
>             if i+direction >= 1 and i+direction <= 4 then
>               for cur_column from -1 to 1 by 2 do
>                 if j-cur_column >=1 and j-cur_column<=4 then
>                   # Determine if the position is free
>                   if A[i+direction, j-cur_column] = 0 then
>                     # Move a single position
>                     Temp:=copy(A);
>                     Temp[i,j]:=0;
>                     Temp[i+direction, j-cur_column]:=piece;
>                     S:=[op(S), copy(Temp)];
>                   elif abs(abs(A[i+direction,j-cur_column])
>                        -piece)=1 then
>                     # We may be able to jump a piece
>                     if (i+2*direction >=1 and
>                          i+2*direction<=4) and
>                        (j-2*cur_column >=1 and
>                          j-2*cur_column<=4) then
>                       # Jump a piece
>                       if A[i+2*direction, j-2*cur_column]
```

```
>                                      = 0 then
>                                      Temp:=copy(A);
>                                      Temp[i,j]:=0;
>                                      Temp[i+direction, j-cur_column]:=0;
>
>                                      Temp[i+2*direction, j-2*cur_column]
>                                                :=piece;
>                                      S:=[op(S), copy(Temp)];
>                                   fi;
>                                fi;
>                             fi;
>                          fi;
>                       od;
>                    fi;
>                 od;
>              if is_king=1 then direction:=-1*direction; fi;
>           fi;
>        od;
>     od;
>     # Check for Kings
>     for i from 1 to nops(S) do
>        for j from 1 to 4 do
>           if S[i][1,j] = 1 then S[i][1,j]:=-1 fi;
>           if S[i][4,j] = 2 then S[i][4,j]:=-2 fi;
>        od;
>     od;
>     # Return list of new board arrangements
>     S;
>  end:
```

To examine this procedure, we will create an initial checkerboard arrangement, called A, using the matrix function of Maple.

```
>  A:=linalg[matrix](4, 4,
>  [[2,0,2,0],[0,0,0,0],[0,0,0,0],[0,1,0,1]]);
```

$$A := \begin{bmatrix} 2 & 0 & 2 & 0 \\ 0 & 0 & 0 & 0 \\ 0 & 0 & 0 & 0 \\ 0 & 1 & 0 & 1 \end{bmatrix}$$

Then, we examine the execution of this procedure on several examples:

```
>  X:=MovePiece(A, 2);
>  Y:=MovePiece(X[1], 1);
>  Z:=MovePiece(Y[3], 2);
```

8.8   Additional Exercises

Exercise 1. Using page 546 of the text as reference, write a Maple procedure for finding the eccentricity of a vertex in an unrooted tree, and for finding the center of an unrooted tree.

Exercise 2. Develop a Maple procedure for constructing rooted Fibonacci trees (see page 547 of the text)

Exercise 3. Develop a Maple procedure for listing the vertices of an ordered rooted tree in level order (see page 604 of the text)

Exercise 4. Construct a Huffman code for the letters of French, based on their occurrence in ordinary French text. The frequency of letters in French is as follows:

E: 18%, A: 8%, I: 7%, U: 6%, O: 5%, Y: 0.2%, S: 8%, N: 8%, T: 7%, R: 7%, L: 6%, D: 4%, C: 3%, M: 3%, P: 2%, V: 2%, F: 1%, Q: 1%, G: 1%, B: 0.9%, H: 0.6%, X: 0.4%, J: 0.3%, Z: 0.06%

Exercise 5. Develop a Maple procedure for producing degree-constrained spanning trees, as outlined on page 604 of the text. Use this procedure on a set of randomly generated graphs to attempt to construct degree-constrained spanning trees where each vertex has degree no larger than 3

Exercise 6. Use Maple to analyze the game of checkers on square boards of different sizes via the technique of game trees

Exercise 7. Develop Maple procedures for finding a path through a maze using the technique of backtracking

Exercise 8. Use Maple to generate as many graceful trees as possible (see page 605 of the text). Can you make any conjectures for this evidence?

Exercise 9. Implement the quick sort in Maple.

Exercise 10. Implement the selection sort in Maple.

Exercise 11. Implement the insertion sort in Maple.

Exercise 12. Use Maple to compare the complexity of different sorting algorithms when sorting the same list of numbers for various initial lists of numbers.

Exercise 13. Alter the postfix expression evaluator to handle prefix expressions.

Exercise 14. Use Maple to animate the steps of different sorting algorithms. Specifically, show each step of the algorithm with one second pauses between movements of elements.

Exercise 15. Modify the code for `Prim` and `Kruskal` so that the edges chosen are displayed as they are selected, and compare the choices of edges.

Exercise 16. Use the procedure provided to count the frequency of occurrence of characters to determine the frequency of characters in various types of English content using relatively large sample input. For example, you might want to use electronic mail messages, computer code, newspaper articles, fiction, and so on.

9 Boolean Algebra

Many situations can be, and are, modeled using a system which can have one of two states. Depending upon the context, the pair of states may be known as "true" and "false", or "on" and "off" or "good" and "bad", and so on. Theoretical examples which spring to mind at once are the precise assertions that are made in mathematical and physical sciences, which are regarded as being either true or false. Among the most important applications is the modeling of digital circuits in computers, in which electronic switches may either be turned on or be turned off.

Now, we have seen how Maple can be used to manipulate arithmetic algebraic systems, and to model their laws, or rules, symbolically. So too can we model the "arithmetic" of boolean algebra. In fact, boolean algebra is somewhat simpler then numerical algebra, so it is actually easier. (At least, it will be once it becomes familiar to you.)

This chapter overlaps to some degree with material covered in Chapter 1, but our emphasis in this chapter will be on circuit minimization and on symbolic aspects of boolean algebra.

We shall discuss the representation of boolean functions in Maple, as well as how to use Maple to verify boolean identities. Next, we show how to use Maple to minimize circuits, represented as boolean expressions. "Don't Care" conditions on circuits are treated, and the techniques we establish for minimization of circuits are extended to deal with these.

9.1 Boolean Functions

In Maple, there is a built in `boolean` type; that is, a Maple variable can be used to represent the boolean values. There are only two boolean values. In Maple, these are represented by the literal constants `true` and `false`. To Maple, these are constant values, just as the number 2, or the 3×3 identity matrix are constants. If you assign either of these two constant values to a variable, then that variable evaluates to that constant value.

```
>  a := true;
```
$$a := true$$

```
>  b := false;
```
$$b := false$$

Two constant values by themselves are not very interesting. To do useful things, we need to be able to operate on them in some meaningful way. Maple provides *two* sets of boolean operators for operating upon boolean variables and literals. The first set consists of the operators **and, or** and **not**. Maple also provides the boolean operators **&and, &or** and **&not** for operating on boolean literals and variables with delayed simplification. They are used in the `logic` package. The difference between the two sets is in the way in which expressions formed using them are simplified.

```
> true and false;
```
$$false$$

```
> true &and false;
```
$$true \ \&and \ false$$

The first **and** operator is the "ordinary" one which yields immediate evaluation, while the second one is more useful for working with boolean expressions symbolically, that is, for studying the form of a boolean expression rather than its value.

A little care is needed in how one uses the **&not** operator. It is important that expressions be sufficiently parenthesized. (First, clear **a** of any value.)

```
> a := 'a';
```
$$a := a$$

While

```
> not not a;
```
$$a$$

is perfectly valid Maple code, the construct

```
> &not &not a;

syntax error, unexpected neutral operator:
&not &not a;
              ^
```

leads, as you can see, to a syntax error. Instead, the latter expression must be typed as

```
> &not (&not a);
```
$$\&not(\&not(a))$$

providing an extra level of parentheses.

The boolean operator **and** is a binary operator that models the semantics of the logical operator "and".

```
>  true and true;
```
> *true*

```
>  true and false;
```
> *false*

```
>  false and true;
```
> *false*

```
>  false and false;
```
> *false*

These four examples exhaust all the possible arguments to the operator **and**. We see that the result of applying **and** to two boolean values is true precisely when both of its operands evaluate to true. Likewise, the example above show that **and** is a commutative operator. In fact, the second and third examples above constitute a proof of this fact.

In a similar way, we can show that the operator **or** is commutative. Its behavior can be determined completely by applying it to all possible pairs of its arguments. (Do this now!)

The operator **not** is a little different than the two binary operator **&and** and **or** in that it is a unary (prefix) operator. Its effect is to "toggle" the two boolean values.

```
>  not true;
```
> *false*

```
>  not false;
```
> *true*

The results are not very surprising!

Maple's boolean operators have a number of other properties of which you should be aware. Both **and** and **or** are *associative* operators. For instance, to say that **and** is a *binary* operator means that it is applied to two arguments at a time. If we wish to compute the "and" of three boolean values, say, a, b, and c, then two distinct expressions may be formed: (a and b) and c as well as a and (b and c). The associative property of the **and** operator asserts that both expressions yield the same value. In fact, once this is established, an inductive argument may be given to show that for any sequence a1, a2, ... , an of boolean values, every way of parenthesizing their conjunction yields the same value. The upshot of this is that parentheses can be dropped. The import of all this discussion is that Maple expressions such as

```
>  a and b and c and d;
```
$$false$$

are entirely unambiguous. Exactly the same thing is true of the operator or.

Another property of which you should be aware is that not is an *involution*. That is,

```
>  not not a;
```
$$a$$

In other words, the second application of not "undoes" the effect of the first.

You can do exactly the same thing with the so-called "inert" variants of these operators (the ones with the ampersand prefixed to their names). However, no automatic simplification will take place. This is because the inert boolean operators are used primarily for working with boolean expressions *symbolically*; they are the operators upon which the symbolic tools in the logic package are based. For example, using the inert operator &not, we see that the expression

```
>  &not( &not ( a ) );
```
$$\&not(\&not(a))$$

is not simplified to just a.

A Boolean Evaluator

Before proceeding further it will be convenient to introduce a new Maple function evalb. This is a general evaluator for boolean expressions. A boolean expression is simple a valid Maple expression constructed from boolean values, variables and operators. However, it is also possible to produce boolean values from other kinds of expressions in Maple, such as arithmetic expressions. For example, the expressions "$2 = 3$" and "$3.14 = 3.14$" are two arithmetic expressions whose values are boolean; that is, either true or false. The function evalb allows us to evaluate expressions such as these as boolean valued expressions in Maple.

```
>  evalb(2 = 3);
```
$$false$$

```
>   evalb(3.14 = 3.14);
```
$$true$$

It is usually from expressions such as these that useful (that is, "practical") boolean values are generated. You have seen this used already, many times, in the test clauses of conditional (if...then...else) and looping (for, while) statements in Maple.

Before we go any further, we should point out that Maple actually understands a *third* boolean value, FAIL. This differs slightly from the discussion in the textbook where only two boolean values are recognized. The value FAIL is a useful addition in a programming language like Maple, however, as it can be used to indicate that a calculation did not complete successfully. Do not confuse FAIL with false; FAIL is a value used to indicate an error in a calculation, not a normal boolean value.

Representing Boolean Functions

Let's look now at how we can represent boolean functions in Maple. These are just like any other function, and can be created using the **proc** command. For example, the boolean function written (using the textbook notation) as

$$F(z, y, z) = xy + yz + zx$$

can be coded as the following Maple procedure.

```
>   F := proc(x, y, z)
>     RETURN((x and y) or (y and z) or (z and x));
>   end:
```

The translation, as you can see, is quite straightforward. A "product" such as xy is translated directly into the Maple expression x and y, while a "sum" $x + y$ is translated as x or y. If you imagine that each "product" xy has an infix dot, such as $x \cdot y$, then the simple rule is to replace each dot by the Maple operator and and replace each "+" by the Maple operator or. The extra parentheses in the definition of F above are not really necessary, but help to improve the readability of the program. (Besides, extra parentheses, correctly placed, never hurt.)

Verifying Boolean Identities

It is a relatively straightforward matter to use Maple to verify boolean identities. For this sort of work, we can use the inert boolean operators. For instance, we can check the distributive law as follows.

```
>   with(logic);  # for 'bequal()'
```
[bequal, bsimp, canon, convert/MOD2, convert/frominert, convert/toinert,

distrib, *dual*, *environ*, *randbool*, *satisfy*, *tautology*]

```
> left := x &or (y &and z);
```
$$left := x \,\&\text{or}\,(y \,\&\text{and}\, z)$$

```
> right := (x &or y) &and (x &or z);
```
$$right := (x \,\&\text{or}\, y) \,\&\text{and}\, (x \,\&\text{or}\, z)$$

```
> bequal(left, right);
```
$$true$$

Here, we have used the Maple library procedure `bequal`, which tests whether or not two boolean expressions are equivalent (returning one of the boolean values `true` and `false` accordingly.) You must have the `logic` package loaded to use this function.

If it turns out that two boolean expressions are not logically equivalent, it may be interesting to determine some assignment of values to the variables which occur in the two expressions which cause the putative identity to fail. The procedure `bequal` can be given an optional third argument into which will be placed such an assignment should the value `false` be returned.

Duality

In Maple, there is a library procedure for finding the dual of a boolean expression. Remember that the dual of a boolean expression is obtained by replacing each occurrence of **and** and **or** by **or** and **and**, respectively. To use it you must load the `logic` package.

```
> with(logic):
```

The procedure is called `dual` (naturally enough), and takes as its argument a boolean expression formed using the inert versions of the boolean operators.

```
> dual(false);
```
$$true$$

```
> dual(true);
```
$$false$$

```
> dual(x &and y);
```
$$x \,\&\text{or}\, y$$

```
> dual(x &or (&not y &or &not x and &not (&not z)));
```
$$x \,\&\text{and}\, (\&\text{not}(y) \,\&\text{and}\, \&\text{not}(x) \,\textbf{and}\, \&\text{not}(\&\text{not}(z)))$$

The beauty of duality is that, once you have proved one boolean identity, you get its dual for free!

While it is possible to use Maple to prove an identity by brute force — by checking every possible value of the variables in it, the logic package offers a much more elegant solution. As an example of this, let's use Maple to prove the identity

$$x\bar{y} + y\bar{z} + z\bar{x} = \bar{x}y + \bar{y}z + \bar{z}x.$$

and formulate our expressions using the inert operators.

```
>   with(logic):  # don't forget to define 'bequal'.
>   left := (x &and &not y)
>        &or (y &and &not z)
>        &or (z &and &not x);
```
$$left := ((x \,\&\text{and} \,\&\text{not}(y)) \,\&\text{or} \,(y \,\&\text{and} \,\&\text{not}(z))) \,\&\text{or} \,(z \,\&\text{and} \,\&\text{not}(x))$$

```
>   right := (&not x &and y)
>        &or (&not y &and z)
>        &or (&not z &and x);
```
$$right := ((\&\text{not}(x) \,\&\text{and} \,y) \,\&\text{or} \,(\&\text{not}(y) \,\&\text{and} \,z)) \,\&\text{or} \,(\&\text{not}(z) \,\&\text{and} \,x)$$

```
>   bequal(left, right);
```
$$true$$

Now we get the dual assertion for free.

```
>   dual(left);
```
$$((x \,\&\text{or} \,\&\text{not}(y)) \,\&\text{and} \,(y \,\&\text{or} \,\&\text{not}(z))) \,\&\text{and} \,(z \,\&\text{or} \,\&\text{not}(x))$$

```
>   dual(right);
```
$$((\&\text{not}(x) \,\&\text{or} \,y) \,\&\text{and} \,(\&\text{not}(y) \,\&\text{or} \,z)) \,\&\text{and} \,(\&\text{not}(z) \,\&\text{or} \,x)$$

```
>   bequal(", "");
```
$$true$$

Disjunctive Normal Form

Maple provides, in its logic package, a function for computing the disjunctive normal form of a boolean expression. It is called canon (short for *canonical*). The following examples typify the calling syntax.

```
>   with(logic):
>   canon((a &or b) &and (c &and d), {a,b,c,d});
```
$$\&\text{and}(c, a, d, false) \,\&\text{or} \,\&\text{and}(c, a, d, \&\text{not}(false))$$

```
> canon((a &or b) &and (&not a &or b), {a,b});
                            false

> canon((a &or b) &and (&not a &or b), {b});

Error, (in logic[distrib]) insufficient variable list
```

The last example shows that there must be at least enough variables to
account for those appearing in the first argument. The first argument to
canon is the expression to be transformed, and the second argument is
the set of variables that are to appear in the disjunctive normal form. It is
possible to specify variables that do not appear in the original equation.

```
> canon(a, {a,b});
        (a &and false) &or (a &and &not(false))
```

In fact, there is a third, optional argument to canon, that specifies *which*
canonical form to produce. Any on of the three values DNF ("disjunctive
normal form",the default), CNF ("conjunctive normal form"), or MOD2,
which directs canon to convert its first argument to an equivalent arithmetic expression modulo 2 (in canonical form).

```
> canon((a &or b) &and (&not a &or &not b), {a,b}, DNF);
        (a &and false) &or (a &and &not(false))

> canon((a &or b) &and (&not a &or &not b), {a,b}, CNF);
            (a &or true) &and (a &or false)

> canon((a &or b) &and (&not a &or &not b), {a,b}, MOD2);
                            a
```

9.2   Representing Boolean Functions

We saw earlier how, given a boolean expression, it is very easy to write
a Maple procedure that is represented by that boolean expression. It is
a simple matter of enveloping the expression in a procedure call. In this
section we will look at what may be considered, in some sense, the opposite
problem. That is, given a boolean function, expressed as a table of values,
how can we find a boolean expression that represents it? Now, we need
first to understand that, since boolean algebra deals with a domain of but
two values, certain simplifications are possible that would not be present
if we were dealing with, say, real-valued functions.

Now we can see why boolean algebra is, in some ways, simpler than some other areas of algebra. In order to specify a boolean valued function f, (whatever its domain might be), it is only necessary to specify which values of its domain are mapped to 1. The rest of the domain of f *must* then be mapped to 0! (Likewise, we could just as well specify the members of the domain of f that are mapped to 0, in which case the remainder would necessarily be those that are mapped to 1.)

This works because the pre-images of the points in the codomain of any function partition the function's domain. For boolean valued functions, there are exactly two sets in the partition, one for each of the boolean values in its codomain. (One of the two sets may be empty.)

This idea is the key to the method outlined in the text for determining a boolean expression representing a given boolean function, and it is the principle upon which we shall base our Maple procedure for computing such expressions.

We are going to write a Maple program that, when fed the pre-image under a boolean function of the value **true**, will compute a boolean expression that represents that function. Later, we shall see a technique for finding a "nice" expression that represents a given function. The procedure we shall write here will compute the so-called "sum of products" representation.

The first step is to design the input to the procedure. As discussed above, it is only necessary to specify the pre-image of (say) **true** (that is, of 1) under our function. This means that we need to input the values as n-tuples which, upon application of our function, yield the value **true**.

Let's consider a fairly simple function, of three variables, with the following table of values.

x	y	z	$f(x, y, z)$
0	0	0	1
0	0	1	0
0	1	0	1
1	0	0	0
0	1	1	1
1	0	1	0
1	1	0	1
1	1	1	1

We should only have to specify which triples (a, b, c) in the domain of f

are mapped to 1. In this case, it is the list

$$(0,0,0),(0,1,0),(0,1,1),(1,1,0),(1,1,1)$$

of triples. So, this is the sort of input we would like to provide to our procedure. Once this is done, it follows that the remaining points

$$(0,0,1),(1,0,0),(1,0,1)$$

in the domain of f are mapped to 0 – it is not necessary to specify this directly.

Now, once we know which n-tuples in the domain of our function are mapped by it to 1, we need to find a minterm corresponding to each of these n-tuples. We'll write this as a subroutine for our main procedure.

```
>   getMinTerm    :=  proc(ll)
>     local   e,  # the minterm, to be returned
>             t,  # temporary variable
>             i;  # loop index
>
>     # check argument
>     if not type(ll, list(boolean)) then
>         ERROR('expecting a list of boolean values');
>     fi;
>
>     # do the construction
>     for i from 1 to nops(ll) do
>         if ll[i] = true then
>             t := 'x'.i;
>         else
>             t := 'not x'.i;
>         fi;
>         if i = 1 then
>             e := t;
>         else
>             e := cat(e, ' and ', t);
>         fi;
>     od;
>     # add parentheses for improved readability
>     RETURN(cat('(', e, ')'));
>   end:
```

This procedure implements the algorithm from the text for computing a minterm to represent a pre-image of 1 (represented in Maple as true). To find the minterm corresponding to the 3-tuple $(1,0,1)$ which, in Maple is written as [true, false, true], we compute

```
>   getMinTerm([true, false, true]);
```
$$(\textit{x1 and not x2 and x3})$$

Having determined minterms for each of the pre-images of 1, all that is left to do is form a "sum" of these minterms. This will be performed by the main procedure for our program.

```
> SumOfProductsExpansion    :=  proc()
>   local   e,  # expression to return
>           i;  # loop variable
>
>   # if no argument, then the
>   # function is identically 'false'
>   if nargs = 0 then
>       RETURN('false');
>   fi;
>
>   for i from 1 to nargs do
>       if not type(args[i], list(boolean)) then
>           ERROR('arguments must be lists of booleans');
>       fi;
>       if i = 1 then
>           e := getMinTerm(args[i]);
>       else
>           e := cat(e, ' or ', getMinTerm(args[i]));
>       fi;
>   od;
>   RETURN(e);
> end:
```

We can use this program to find a boolean expression representing our example function as follows.

```
> SumOfProductsExpansion(
>   [false, false, false],
>   [false, true, false],
>   [false, true, true],
>   [true, true, false],
>   [true, true, true]
> );
```
$(not\ x1\ and\ not\ x2\ and\ not\ x3)\ or\ (not\ x1\ and\ x2\ and\ not\ x3)\ or\ (no\backslash$
$t\ x1\ and\ x2\ and\ x3)\ or\ (x1\ and\ x2\ and\ not\ x3)\ or\ (x1\ and\ x2\ and\ x3\backslash$
$)$

9.3 Minimization of Boolean Expressions and Circuits

Our function for finding the sum of products expansion of a boolean function may lead to inefficient circuits, because the number of gates required to implement it directly may very well be greater than the number that is really necessary. Although the number of distinct boolean functions of

n variables (where n is a positive integer) is finite — in fact, it is equal to 2^{2^n}, as shown in the text — it is easy to see that the number of distinct boolean *expressions* on n variables is infinite. Some form of the Pigeon Hole Principle compels us to conclude that some boolean function has many — indeed, infinitely many — distinct representations by boolean expressions.

From the perspective of circuit design, therefore, what is needed is a method for "minimizing" a circuit, in the sense that we would like, given a circuit, to find an equivalent circuit that uses as few gates as possible.

To make Maple do this work for us, we must translate the problem from the pictorial language of circuit diagrams to an algebraic description involving boolean expressions, recognizing that a circuit diagram is a pictorial representation of an equivalent boolean expression, wherein a logic gate simply represents one of the standard boolean operators: **and**, **or** and **not**.

To make this a bit more concrete, let's look at a simple example. The boolean expression

$$xy + x\bar{y}$$

which, rendered in the syntax of Maple, looks like

```
>  e := (x &and y) &or (x &and &not y);
           e := (x &and y) &or (x &and &not(y))
```

can be "minimized" by using Maple as follows.

```
>  with(logic):
>  distrib(x &and (y &or &not y));
           (x &and y) &or (x &and &not(y))
```

which shows that the expressions

$$x(y + \bar{y}) \qquad \text{and} \qquad xy + x\bar{y}$$

represent the same boolean function. But $y + \bar{y} = 1$, for any y:

```
>  bequal(y &or &not y, true);
                    true
```

so that $x(y + \bar{y})$ simplifies further to x.

The trick is to spot, for a given boolean expression, opportunities to eliminate variables, or reduce the number of minterms, by using algebraic properties of boolean operators. For the simple example above, this was very easy, and Maple merely allowed us to prove that our guesses were correct. But expressions that are just a little more complicated may require a great deal more thought to spot such simplifications. What we need is something that will allow us to work in the opposite direction to that taken above; namely, given the original expression, can we actually *find* a simpler expression to which it is equivalent? Further, can we find one which is minimal?

Fortunately, Maple's `logic` package provides a "circuit minimizer" that takes care of all of this for us. It is called `bsimp`. To use this procedure, you must load the `logic` package first into your Maple session, or else call it by its long form name `logic[bsimp]`.

Of course, Maple does not speak directly in terms of gates and circuits. To ask Maple to minimize a circuit, you must speak Maple's algebraic language by specifying the equivalent boolean expression.

For instance, to simplify our earlier example

```
>   e := (x &and y) &or (x &and &not y);
```
$$e := (x \,\&\text{and}\, y) \,\&\text{or}\, (x \,\&\text{and}\, \&\text{not}(y))$$

you can type

```
>   with(logic):  # load 'bsimp'
>   bsimp(e);
```
$$x$$

You can apply `bsimp` to any boolean expression formed using the inert boolean operators from the `logic` package. Let's see how Maple handles some more complicated examples.

```
>   with(logic):
>   e := (w &and x &and y &and (&not z))
>     &or (w &and (&not x) &and y &and z)
>     &or (w &and (&not x) &and y &and (&not z))
>     &or ((&not w) &and x (&not y) &and z)
>     &or ((&not w) &and (&not x) &and y &and z)
>     &or ((&not w) &and (&not x) &and (&not y) &and z);
```

$$e := ((((((w \,\&\text{and}\, x) \,\&\text{and}\, y) \,\&\text{and}\, \&\text{not}(z)) \,\&\text{or}$$
$$(((w \,\&\text{and}\, \&\text{not}(x)) \,\&\text{and}\, y) \,\&\text{and}\, z)) \&\text{or}$$
$$(((w \,\&\text{and}\, \&\text{not}(x)) \,\&\text{and}\, y) \,\&\text{and}\, \&\text{not}(z))) \&\text{or}$$
$$((\&\text{not}(w) \,\&\text{and}\, x(\&\text{not}(y))) \,\&\text{and}\, z)) \&\text{or}$$
$$(((\&\text{not}(w) \,\&\text{and}\, \&\text{not}(x)) \,\&\text{and}\, y) \,\&\text{and}\, z)) \&\text{or}$$
$$(((\&\text{not}(w) \,\&\text{and}\, \&\text{not}(x)) \,\&\text{and}\, \&\text{not}(y)) \,\&\text{and}\, z)$$

```
>   bsimp(e);
```

The `bsimp` procedure is very complex, and uses a Quine-McCluskey algorithm based upon representing boolean expressions as sets. These data structures, while very natural for Maple, do not correspond very well with the description given in the text.

9.4   Don't Care Conditions

It is possible to use our procedure `SumOfProductsExpansion` to handle the so-called "don't care conditions" discussed in your textbook. Here, we'll develop a Maple procedure that allows us to compute a sum of products expansion (disjunctive normal form), of minimum length, for a boolean function specified together with don't care conditions.

Informally, a set of don't care conditions for a boolean function f is a set of points in the domain of f with whose images under f we are not concerned. In other words, we don't care where f sends those points. There are likely some points in the domain of f whose images under f are important to us however. These points at which f is well-defined form a subset A of the domain of f, and the restriction of f to A is a well-defined function (in the sense that there is no ambiguity over the value of f at any of these points). Note that, if f is to be a function of n variables, then the domain D is simply the set $\{0,1\}^n$ of all n-tuples of 0s and 1s (or, in Maple syntax, the set $\{true, false\}$).

If we think of f as a fully defined function on this subset A of D, then what we are interested in is the family of all extensions of f to D. That is the set of all boolean valued functions g on D whose restriction to A is equal to f. Now, each of these functions g is completely defined on D, so the technique we used earlier to compute sum of products expansions can be applied to any one of them. So, to find an "optimal" (which, here, means "smallest") sum of products expansion of "f", we can compute the unique sum of products expansion for each extension g of f to D, and search among them for one of minimum size.

We should pause to consider the size of this problem. The subset A of D upon which f is well-defined, and the subset DC of "don't care points", upon which f is not specified partition the domain D. That is, D may be written as a disjoint union

$$D = A \cup DC.$$

If there are d don't care points (that is, if $|D| = d$), then there are 2^d

extensions g of f to D. Each such extension corresponds to one choice of
a subset of DC to include among the pre-images of 1. So, the problem
grows very quickly with the number of don't care points, or conditions.

The optic that we have adopted here makes it very easy to see an algorithm that computes a sum of products expansion, of minimum length,
for a function given with don't care conditions. We can simply use an
exhaustive search over the set of all well defined extensions.

To do this, we will write a Maple procedure dcmin ("Don't Care
Minimizer") to construct all of the 2^d functions, call our procedure
SumOfProductsExpansion on each, and then look for one that has mini-
mum length. Here is the Maple code to do this.

```
>   dcmin :=  proc(pt::{set,list}, dc::{set,list})
>     local  e,  # expression to return
>          te, # temporary expression
>          i,  # index
>          s,  # the size of the smallest expression so far
>          PT, # pt as a set
>          DC, # dc as a set
>          PDC,    # power set of DC
>          T,  # temporary set (loop variable)
>          S;  # temporary domain for well-defined functions
>
>     PT := {op(pt)};
>     DC := {op(dc)};
>     PDC := combinat[powerset](DC);
>
>     s := infinity;
>
>     for T in PDC do
>          S := T union PT;
>          te := SumOfProductsExpansion(op(S));
>          if evalb(length(te) < s) then
>              e := te;
>              s := length(e);
>          fi;
>     od;
>     if s = infinity then
>          ERROR('can't happen');
>     else
>          RETURN(e);
>     fi;
>   end:
```

This is simply a brute force solution to the problem. We loop over
all possible sets of values in the "don't care" set DC which, in effect, allows us to specify a unique well defined function for input to
SumOfProductsExpansion. We examine the length of the expression te
returned for each, and, should it prove to be smaller than any other ex-

pression seen so far, we record it as the new value of **e**. When all the possibilities have been exhausted, the variable **e** will contain an expression of shortest length representing the input function.

Note that there may, in fact, be several expressions whose length is equal to this minimum value. Our procedure returns the first one that it encounters.

We have made this procedure a little more user friendly by designing it so that it accepts either a pair of lists or a pair of sets for input.

Note that, as adopted here, the length of an expression is just one measure of its complexity. You may, for example, wish to count the number of boolean operators in the expression and minimize that number. You could change this procedure simply by replacing the function **length** by some other measure of an expressions complexity.

Now that we have a procedure for minimization Boolean functions, let's try it out with some examples Consider a boolean function f with the following table of values, in which a "d" in the right most column indicates a "don't care" condition.

x	y	z	$f(x, y, z)$
false	false	false	true
false	false	true	false
false	true	false	d
true	false	false	d
false	true	true	true
true	false	true	false
true	true	false	false
true	true	true	true

The input consists of the set

$$\{[\texttt{false, false, false}], [\texttt{false, true, true}], [\texttt{true, true, true}]\}$$

of points mapping to **true** (the pre-image **pt** of **true**), and the set

$$\{[\texttt{false, true, false}], [\texttt{true, false, false}]\}$$

of points that we don't care about. We can compute a sum of products expansion for this function as follows.

```
>   dcmin(
>     {[false, false, false],
>       [false, true, true],
>       [true, true, true]},
```

```
>   {[false, true, false],
>    [true, false, false]});
```
(*not x1 and not x2 and not x3*) *or* (*not x1 and x2 and x3*) *or* (*x1 and*\
    x2 and x3)

As we mentioned above, we could just as well have represented the input as two lists:

```
> dcmin(
>    [[false, false, false],
>     [false, true, true],
>     [true, true, true]],
>    [[false, true, false],
>     [true, false, false]]);
```
(*not x1 and not x2 and not x3*) *or* (*not x1 and x2 and x3*) *or* (*x1 and*\
    x2 and x3)

(The former is more readable, while the latter is more consistent with the input to SumOfProductsExpansion.)

9.5 Computations and Explorations

In this section, we shall look at Problems 2 and 6 from the *Computations and Explorations* section of the text, and see how we can use Maple to solve some of them.

2 Construct a table of the boolean functions of degree 3.

Solution: First, we should note that there are
```
>   2^(2^3);
```
$$256$$

boolean functions of degree 3, so the output of our computation is going to be lengthy. We can use the function SumOfProductsExpansion that we developed earlier to help us with this calculation. Remember that the sum of products expansion, or disjunctive normal form, of a boolean function provides a bijective correspondence between boolean functions, and certain boolean expressions. Although there are infinitely many boolean expressions, there are precisely 2^{2^n} disjunctive normal forms on n variables, and they are in bijective correspondence with the set of all boolean functions of n variables, so this is a convenient representation to use.

To generate the entire list, we need to specify all possible (distinct) argument lists to `SumOfProductsExpansion`. This means that we must generate the power set of the domain of all boolean functions of three variables. Now, a boolean function f of three boolean variables has domain equal to

```
>   Dom3 := {[false,false,false],
>           [false,false,true],
>           [false,true,false],
>           [true,false,false],
>           [true,false,true],
>           [true,true,false],
>           [false,true,true],
>           [true,true,true]};
```

$Dom3 := \{[true, false, true], [false, false, false], [false, true, false],$
$\quad [false, true, true], [true, true, false], [true, true, true], [true, false, false],$
$\quad [false, false, true]\}$

We can generate its power set by using the `powerset` procedure in the `combinat` package. Therefore, we must first load the `combinat` package.

```
>   with(combinat):
```

```
>   PDom3 := combinat[powerset](Dom3);
```

Notice the form of the output. We obtain a set of sets, while our `SumOfProductsExpansion` procedure requires an expression sequence of lists of booleans. This means that we need to use Maple's `op` function as well. We can now generate the list of boolean functions in three variables using a simple `for` loop:

```
>   for S in PDom3 do
>       print(SumOfProductsExpansion(op(S)));
>   od;
```

$$false$$

$(not\ x1\ and\ not\ x2\ and\ not\ x3)\ or\ (not\ x1\ and\ x2\ and\ x3)\ or\ (x1\ and\backslash$
$\quad x2\ and\ x3)$
$\quad\quad (not\ x1\ and\ x2\ and\ not\ x3)\ or\ (x1\ and\ not\ x2\ and\ not\ x3)$
$$(not\ x1\ and\ x2\ and\ not\ x3)$$
$$(x1\ and\ not\ x2\ and\ not\ x3)$$

$$...$$

$$much\ output\ deleted$$

$$...$$

*(not x1 and not x2 and not x3) or (not x1 and x2 and x3) or (x1 and\
   x2 and not x3) or (x1 and x2 and x3) or (x1 and not x2 and not x3\
)*
*(x1 and not x2 and x3) or (not x1 and not x2 and not x3) or (not x1 \
   and x2 and x3) or (x1 and x2 and not x3) or (x1 and x2 and x3) or \
   (x1 and not x2 and not x3)*
 *(not x1 and x2 and not x3) or (not x1 and x2 and x3) or (x1 and x2 \
   and not x3) or (x1 and x2 and x3) or (x1 and not x2 and not x3)*
*(x1 and not x2 and x3) or (not x1 and x2 and not x3) or (not x1 and\
   x2 and x3) or (x1 and x2 and not x3) or (x1 and x2 and x3) or (x1\
   and not x2 and not x3)*
 *(not x1 and not x2 and not x3) or (not x1 and x2 and not x3) or (no\
   t x1 and x2 and x3) or (x1 and x2 and not x3) or (x1 and x2 and x3\
) or (x1 and not x2 and not x3)*
*(x1 and not x2 and x3) or (not x1 and not x2 and not x3) or (not x1 \
   and x2 and not x3) or (not x1 and x2 and x3) or (x1 and x2 and no\
   t x3) or (x1 and x2 and x3) or (x1 and not x2 and not x3)*

You might want to write the output to a file to better be able to examine it. You can do this by using the `printf` function in place of `print`. Alternatively, you may use the `writeto` or `appendto` functions to redirectMaple output to a file. Once you are done, you can restore terminal output by issuing the call

```
>   writeto(terminal);
```

Use Maple's `help` facility to learn more about these functions if necessary.

6 Randomly generate ten boolean expressions of degree 4, and determine the average number of steps to minimize them.

Solution: There are really two parts to this problem: First, we need to find a way to generate random boolean expressions. Second, we need to find some method of examining the minimization process so that we can count the steps.

Mapleprovides an easy solution to the first part of the problem. In the `logic` package, there is a procedure called `randbool` that generates a random boolean expression. Since it is part of the `logic` package, it must be defined before it can be used.

```
>  with(logic):
```

To use `randbool`, you must specify an alphabet upon which to construct boolean expressions. For example, to generate a random boolean expression on the symbols a and b, you can type

```
>  randbool({a,b});
```
$$\&\text{or}(a \text{ \&and } \&\text{not}(\text{false}), \&\text{not}(a) \text{ \&and } \&\text{not}(\text{false}), \text{false \&and } a)$$

or

```
>  randbool([a,b]);
```
$$a \text{ \&and } \&\text{not}(\text{false})$$

That is, the alphabet can be specified either as a set or as a list. Notice that the results of the two calls above are different — the expressions are generated *randomly*. (This is true even after a `restart`. You can try this too, but don't forget to load the `logic` package again.) Now, to generate ten random boolean expressions, we can simply use a for loop:

```
>  for i from 1 to 10 do
>     randbool({a,b});
>  od;
```
$$\&\text{or}(a \text{ \&and } \&\text{not}(\text{false}), \&\text{not}(a) \text{ \&and } \&\text{not}(\text{false}), \text{false \&and } a,$$
$$\text{false \&and } \&\text{not}(a))$$
$$(\&\text{not}(a) \text{ \&and } \&\text{not}(\text{false})) \text{ \&or } (\text{false \&and } a)$$
$$\&\text{or}(a \text{ \&and } \&\text{not}(\text{false}), \text{false \&and } a, \text{false \&and } \&\text{not}(a))$$
$$\text{false \&and } a$$
$$(\&\text{not}(a) \text{ \&and } \&\text{not}(\text{false})) \text{ \&or } (\text{false \&and } \&\text{not}(a))$$
$$\&\text{or}(a \text{ \&and } \&\text{not}(\text{false}), \&\text{not}(a) \text{ \&and } \&\text{not}(\text{false}), \text{false \&and } a)$$
$$(\text{false \&and } a) \text{ \&or } (\text{false \&and } \&\text{not}(a))$$
$$(a \text{ \&and } \&\text{not}(\text{false})) \text{ \&or } (\text{false \&and } a)$$
$$(a \text{ \&and } \&\text{not}(\text{false})) \text{ \&or } (\&\text{not}(a) \text{ \&and } \&\text{not}(\text{false}))$$
$$(a \text{ \&and } \&\text{not}(\text{false})) \text{ \&or } (\text{false \&and } a)$$

Notice also that the expressions generated are in disjunctive normal form, that is, a sum of products expansion. It is also possible to specify a second argument to `randbool` that affects the form of the generated expressions. The second argument can have any one among the values `DNF`, `CNF` or `MOD2`, just like the `canon` procedure we met earlier.

```
>  randbool({x,y,z}, CNF);
```
$$\&\text{and}(\&\text{or}(y, z, \&\text{not}(x)), \&\text{or}(\&\text{not}(y), \&\text{not}(x), \&\text{not}(z)),$$
$$\&\text{or}(x, \&\text{not}(y), \&\text{not}(z)), \&\text{or}(x, z, \&\text{not}(y)), \&\text{or}(x, y, z),$$
$$\&\text{or}(z, \&\text{not}(y), \&\text{not}(x)), \&\text{or}(y, \&\text{not}(x), \&\text{not}(z)))$$

```
>  randbool({u,v}, MOD2);
```
$$1 + v + u + v\,u$$

Having solved the first part of the problem, we need to find a way to count the number of steps taken during the minimization process. There are three approaches that we can take to this part of the problem.

The first is to to measure the time taken to execute a procedure. You have seen this before in previous chapters.

The second is the `trace` facility to monitor the number of steps taken to perform a minimization. You can trace a Maple procedure by issuing the call

```
>   readlib(trace):
>   trace(bsimp);
```

Now if we generate a random boolean expression `e` on four variables

```
>   e := randbool({a,b,c,d});
```
$e := \&or(\&and(c, d, \&not(a), \&not(\mathit{false})), \&and(\mathit{false}, c, a, \&not(d)),$
$\qquad \&and(\mathit{false}, a, \&not(c), \&not(d)), \&and(\mathit{false}, c, a, d),$
$\qquad \&and(\mathit{false}, a, d, \&not(c)), \&and(a, d, \&not(c), \&not(\mathit{false})),$
$\qquad \&and(a, \&not(c), \&not(d), \&not(\mathit{false})),$
$\qquad \&and(\mathit{false}, c, \&not(a), \&not(d)), \&and(c, \&not(a), \&not(d), \&not(\mathit{false})))$

we can observe the number of steps required to simplify it by simply calling `bsimp` after tracing it.

```
>   bsimp(e):
```

The tracing of `bsimp` will print a number of statements of the form `B :=` `<something>`, which you can count. We have not shown them here.

The `trace` procedure actually causes the execution of `bsimp` to print out more information than we need. Some of this information can be ignored for this problem. Each executed statement is printed, as are the arguments and return values of any subroutines called. We really only want to count the number of statements executed, so the other information can simply be ignored.

To undo the effect of the `trace` procedure, you can "untrace" a function by calling, naturally enough, the `untrace` procedure:

```
>   untrace(bsimp); # 'bsimp' will no longer be traced
```
$$bsimp$$

Finally, to get the most information, we can set the `printlevel` variable. The `printlevel` variable is a bit confused, in that it does not know whether it is a global or a local variable. Like any global variable, it is visible at the Maple top level. However, its value is changed every time it enters a control structure, such as a loop, or a procedure body,

and then reset to its original value after leaving the structure. Normally,
`printlevel` is set to 1.

```
> printlevel;
```
$$1$$

But it is possible to assign a value to printlevel, which will affect how
much is printed inside control structures.

```
> for i from 1 to 5 do
>    sqrt(i);
> od;
```
$$1$$
$$\sqrt{2}$$
$$\sqrt{3}$$
$$2$$
$$\sqrt{5}$$

Try the preceding loop after setting the value of `printlevel` to something
like 16. Each nested level of procedure invocation decrements `printlevel`
by 5. Thus, setting `printlevel` to 6 at the top level is very much like
tracing (using `trace`) all Maple procedures.

To get Maple to display what it is doing at even further levels of function
calls, simple set `printlevel` to a high value.

```
> printlevel := 16;

> for i from 1 to 5 do
>    sqrt(sin(abs(i)));
> od;
```
$$\sqrt{\sin(1)}$$
$$\sqrt{\sin(2)}$$
$$\sqrt{\sin(3)}$$
$$i\sqrt{-\sin(4)}$$
$$i\sqrt{-\sin(5)}$$

For typographical reasons, the voluminous output has been suppressed
in this printed manual, but you can see the spectacular results on your
computer screen. Try doing this last example with `printlevel` set to a
very large value like 10000.

Now, to really see what is happening when you invoke `bsimp` to minimize
boolean expression, you can set `printlevel` to a huge value, and observe
the results.

```
> with(logic):  # make sure that 'bsimp' is defined

> printlevel := 100000;

> e := x &and y &or &not z;
```
$$e := (x \mathbin{\&and} y) \mathbin{\&or} \mathbin{\&not}(z)$$

```
>  bsimp(e);
```

$$\&\mathrm{not}(z)\,\&\mathrm{or}\,(x\,\&\mathrm{and}\,y)$$

To interpret the results of the output, it is helpful to be able to look at the source code for the **bsimp** procedure, and the library subroutines that it calls. You can do this by issuing the following

```
>  interface(verboseproc = 2);
>  eval(bsimp);  # assumes 'bsimp' is loaded
```

proc(b)
    local B, u;
    option '*Copyright 1990 by the University of Waterloo*';
        if not readlib('*logic/isboolean*')(b, 'u') **then**
            ERROR('*undefined logical operators*', u)
        fi;
        $B := b$;
        $B := logic_{distrib}(B)$;
        $B := $ 'logic/primeimp'(B);
        $B := $ 'logic/irred'(B)
    end

This shows the actual source code for the **bsimp** function, although comments are missing. (There can be no comments because Maple "disassembles" the object code in the Maple library, from which compilation has stripped all of the comments, to produce the Maple code you see here.)

To understand the output, you should realize that Maple does not provide access to bit level operations, so the algorithm used in **bsimp** is somewhat different than the one described in your textbook. Instead of bit strings, Maple uses sets to represent boolean expressions.

With these tools in hand, you can now write a procedure to generate, randomly, ten boolean expressions on four variables, and count the number of steps needed to minimize each, finally taking an average.

9.6   Exercises/Projects

Exercise 1. Use Maple to verify DeMorgan's Laws and the commutative and associative laws. (See Page 612 of the text; Table 5.)

Exercise 2. Use Maple to construct truth tables for each of the following pairs of boolean expressions. Hence, or otherwise, decide upon their logical equivalence.

- $a \rightarrow b$ and $b \rightarrow a$
- $a \rightarrow \bar{b}$ and $b \rightarrow \bar{a}$
- $a + (b\bar{c})$ and $(a + b + d)(a + c + d)$

Exercise 3. Write a Maple procedure that constructs a table of values of a boolean expression in n variables that may include the following operators: &and, &or, &xor, &nand, &nor.

Exercise 4. Write a Maple procedure that given a boolean function represents this function using only the &nand operator.

Exercise 5. Use the procedure in the previous exercise to represent the following boolean functions using only the nand operator.

(a) $F(x, y, z) = xy + \bar{y}z$

(b) $G(x, y, z) = x + \bar{x}y + \bar{y}\bar{z}$

(c) $H(x, y, z) = xyz + \bar{x}\bar{y}\bar{z}$

Exercise 6. Write a Maple procedure that given a boolean function represents this function using only the &nor operator.

Exercise 7. Use the procedure in the previous exercise to represent the boolean functions in Exercise FIXME using only the &nor operator.

Exercise 8. Write a Maple procedure for determining the output of a threshold gate, given the values of n boolean variables as input, and given the threshold value and a set of weights for the threshold gate. (See Page 648 of the text.)

Exercise 9. Develop a Maple procedure that, given a boolean function in four variables, determines whether it is a threshold function, finding the appropriate threshold gate representing this function. (See Page 648 of the text.)

Exercise 10. A boolean expression e is called **self dual** if it is logically equivalent to its dual e^d. Write a Maple procedure to test whether a given expression is self dual.

Exercise 11. Determine, for each integer $n \in \{1, 2, 3, 4, 5, 6\}$ the total number of boolean functions of n variables, and the number of those functions that are self dual.

Exercise 12. Write a function `ProductOfSumExpansion` that operates like our `SumOfProductsExpansion`, taking the same input, but computing a boolean expression in the form of a "product of sums" representing the input function.

Exercise 13. Our procedure `SumOfProductsExpansion` has a misfeature. Imagine a boolean function of 1996 variables for which $f(0,0,0,\ldots,0) = 0$, and for which all the other $2^{1996} - 1$ points in the domain of f are mapped to 1. You must type in all the other $2^{1996} - 1$ points, because it is those points that are expected as input. Obviously, this is not humanly possible. Rewrite `SumOfProductsExpansion` to fix this problem.

Exercise 14. Write a Maple procedure that, given a positive integer n, constructs a list of all boolean functions of degree n. Use your procedure to find all boolean functions of degree 4.

Exercise 15. Use `dcmin` to compute a minimal sum of products expansion for a boolean function with don't care conditions specified by the following Karnaugh map, in which a "d" indicates a "don't care" condition.

	$y\,z$	$\bar{y}\,z$	$y\,\bar{z}$	$\overline{y\,z}$
x	1	d	d	1
$\bar{x}$			1	

Exercise 16. How can you change exactly one character in the definition of the procedure `dcmin` so that it returns the *last* expression of minimum length representing the input function that it encounters. (As written now, it returns the *first*; see the discussion following its implementation.)

Exercise 17. Because of the way that Maple is implemented, our function `dcmin`, as designed here, may yield different output in different Maple sessions for the same input. To understand this, you need to realize that, during one Maple session, the set $\{a, b, c, d\}$ may be represented by Maple as `a,b,c,d` and, in another session, it may appear as `c,b,d,a`. Redesign the function so that this does not occur. To do this, you will need to consult some of the Maple reference documentation to learn about how Maple stores set in memory. (**Hint:** Consider the `type` of the input and, if necessary, read about how Maple stores objects in memory.)

Exercise 18. Do Question 4. Now write a Maple procedure to generate random boolean expressions in 4 variables, and stop when it has found one that is self dual. Run the program several times and time it, taking an average. Now do the same thing for boolean expressions in 5 and 6 variables. Generalize.

Exercise 19. Investigate the function `convert` using Maple's "help" facility. Use it to write versions of `bsimp` and `bequal` that accept boolean expressions as input that are written in the arithmetic notation used in your textbook. (*Hint:* **Don't** rewrite these two functions completely. Use them to help you.)

Exercise 20. Revise the procedure `dcmin` to return *all* minimal expressions that it finds, rather than just the first.

Exercise 21. Revise the procedure `dcmin` to use different measures of complexity of boolean expressions, such as the number of boolean operations, etc.

10   Modeling Computation

10.1   Introduction

This chapter describes how to use Maple to work with the computational models discussed in Chapter 10 of the text, including finite-state machines, regular expressions, and Turing machines. Although Maple does not have direct support for these structures, it provides a rich programming environment that can be used to develop procedures that simulate the behavior of computational models.

In this chapter we will develop Maple procedures that can be used to study various computational models, including finite-state machines and Turing machines. In particular, we will use the structures provided by Maple, including sets, lists, sequences, and tables to write these Maple procedures.

Before discussing how to use Maple to simulate computational machines, we need to introduce a simple data structure called a stack. After covering stacks, we will describe how to implement different types of finite automata. We will implement deterministic finite-state automata (DFA), non-deterministic finite-state automata (NFA), and finite-state machines with output. After covering these different types of machines, we will discuss a routine to convert an NFA to a DFA. We will also describe how to construct a finite-state machine that recognizes a given regular expression.

The code described in this chapter can be found at ftp.maplesoft.com. It would be prudent to obtain the latest release of the source code and README file, in case new features of bug fixes are provided.

After discussing finite-state machines we will turn our attention to Turing machines. We will implement Turing machines using techniques developed for finite-state machines as a basis, adding the tape included in a Turing machine. We will use Maple to study busy beaver machines, which are Turing machines that produce the largest number of ones on an initially blank tape, given that the Turing machine has n states and has alphabet 1,B.

Step	Operation	Stack
1	push(1)	1
2	push(2)	21
3	push(3)	321
4	push(4)	4321
5	top()	4321
6	pop()	321
7	pop()	21
8	push(5)	521
9	pop()	21
10	pop()	1
11	pop()	
12	pop()	*ERROR*

Table 10.1: Table An example of a stack through various stack operations.

10.2 Stacks

Some of our routines for finite-state machines will make use of a data structure known as a stack. A stack is a dynamic data structure with three operations: push, top, and pop. A stack has a "top" element, whose value is returned with the top operation, and which is removed from the stack during a pop operation. A push operation places a new top element on the stack.

We can draw a stack on a single line of text as a sequence with the top element as the leftmost symbol on the line. An example of a sequence of stack operations is shown in Table 10.1. Notice that in step 12, there is no element to pop resulting in the output: ERROR.

Maple provides a data structure, called a `list`, which allows us to order elements such that the first element of the list is the top of the stack. A list is created by enclosing Maple objects in square brackets, as in

```
>  list1 := [1, 2, 3];
```
$$list1 := [1, 2, 3]$$

```
>  whattype(list1);
```
list

Since the order of elements in a list matters, [1, 2, 3] does not equal [2, 3, 1]. Note that we can obtain the number of elements in a list with the Maple nops function, and we can extract the ith element of a list, as well as the entire list, with the op function:

```
>  nops(list1);
```
$$3$$

```
>  op(2, list1);
```
$$2$$

```
>  op(list1);
```
$$1, 2, 3$$

In addition to the op function, we can access elements and subranges of a list *list1* by including a range between square brackets, as in

```
>  list1[2..3];
```
$$[2, 3]$$

```
>  whattype(list1[2..3]);
```
$$list$$

For Maple V Release 3, the object returned from $list1[2..3]$ is an expression sequence rather than another list. However, we can create a list from a sequence by enclosing the sequence in square brackets, as in

```
>  [ list1[2..3] ];
```
$$[[2, 3]]$$

```
>  whattype( [ list1[2..3] ] );
```
$$list$$

For Maple V Release 4, list1[2..3] will return a list rather than an expression sequence.

We are now ready to discuss how to simulate a stack using Maple lists. The top operation first checks that there is an element in the list; if there is, the op function is carried out to extract the first element:

```
>  StackTop := proc(stack)
>      if nops(stack) = 0 then
>          ERROR('Stack Top with Empty Stack');
>      fi;
>      RETURN(op(1, stack));
>  end:
```

The pop operation is very similar, but returns a new list with the first element, i.e. the top of the stack, removed.

```
>   StackPop := proc(stack)
>       if nops(stack) = 0 then
>           ERROR('Stack Pop with Empty Stack');
>       fi;
>       RETURN(RI_Sublist(stack, 2, nops(stack)));
>   end:
```

The routine RI_SubList returns a sublist as a list for both Maple V Release 3 and Maple V Release 4 after detecting which version you are currently running.

Finally, the push operation returns a new stack with the new top element added to the front, or left-hand side, of the list. (Unlike the top and pop operations, a push operation does not require a check that there is an element on the stack.)

```
>   StackPush := proc(stack, newtop)
>       RETURN([newtop, op(stack)]);
>   end:
```

We finish off our discussion about stacks by showing the Maple code for the example in Table 10.1.

```
>   stack:=[];
```
$$stack := []$$

```
>   stack := StackPush(stack, 1);
```
$$stack := [1]$$

```
>   stack := StackPush(stack, 2);
```
$$stack := [2, 1]$$

```
>   stack := StackPush(stack, 3);
```
$$stack := [3, 2, 1]$$

```
>   stack := StackPush(stack, 4);
```
$$stack := [4, 3, 2, 1]$$

```
>   X := StackTop(stack);
```
$$X := 4$$

```
>   stack := StackPop(stack);
```
$$stack := [3, 2, 1]$$

```
>   stack := StackPop(stack);
```
$$stack := [2, 1]$$

```
>   stack := StackPush(stack, 5);
```
$$stack := [5, 2, 1]$$

```
>   stack := StackPop(stack);
```
$$stack := [2, 1]$$

```
>   stack := StackPop(stack);
```
$$stack := [1]$$

```
>   stack := StackPop(stack);
```
$$stack := []$$

```
>   stack := StackPop(stack);
```

```
Error, (in StackPop) Stack Pop with Empty Stack
```

10.3 Finite-State Machines with Output

We are now ready to discuss our first computational model, finite-state machines with output. These are machines which generate an output symbol for each (state, input symbol) pair observed during the simulation of the machine on a list of input symbols. The concatenation of these generated symbols constitutes the output of the machine for the input list. A finite-state machine is defined by a set of states, input and output alphabets, a transition function, an output function, and a start symbol. We must first be able to represent these items in Maple. Then, we can use Maple routines to simulate a finite-state machine on an input list.

Our representation allows any Maple object to be used as an alphabet symbol or a state. We can therefore use integers, strings, sets, or lists as either states or alphabet symbols. The set of states is represented by a Maple set. We also represent the input and output alphabets using Maple sets. For the transition function and output function, we have chosen Maple tables. A table in Maple can be used to map lists of Maple objects to other Maple objects, such as lists. Each mapping of a list of Maple objects onto another Maple object is considered to be an entry in the table. The lists in the domain of the table are called the indices. The elements of the range are called entries. The number of items in a list of the domain can vary. For example, we can define a table on the list [a, b, c] as well as on [a, b, d, g]. We will see later how this flexibility can be exploited to represent non-deterministic finite-state machines.

Now that we have representations for each of the parts of a finite-state machine with output, we can put them all together in a Maple list. To summarize, we represent a finite-state machine $M = (S, I, O, f, g, s_0)$ in Maple by a list of six items:

Item	Type	Description
1	Set	The set of states, S, for M
2	Set	The set of input alphabet symbols, I
3	Set	The set of output alphabet symbols, O
4	Table	The transition function, f
5	Table	The output function, g
6	any type	The starting symbol, s_0

As an example, we will give the representation for the finite-state machine used in Example 4 on page 667 of the text. Notice that entries in a Maple table *Delta* are defined by placing the domain elements in square brackets and assigning the range element. The statement "Trans := table();" is important. It makes sure that there are no previous entries for a table called Trans.

```
>   States := { s0, s1, s2, s3, s4 };
                States := {s0, s1, s2, s4, s3}

>   InputAlphabet   := {0, 1};
                InputAlphabet := {0, 1}

>   OutputAlphabet := {0, 1};
                OutputAlphabet := {0, 1}

>   Trans := table():      # clear Trans
>   Trans[s0, 0] := s1: Trans[s0, 1] := s3:
>   Trans[s1, 0] := s1: Trans[s1, 1] := s2:
>   Trans[s2, 0] := s3: Trans[s2, 1] := s4:
>   Trans[s3, 0] := s1: Trans[s3, 1] := s0:
>   Trans[s4, 0] := s3: Trans[s4, 1] := s4:
>   Output := table():
>   Output[s0, 0] := 1: Output[s0, 1] := 0:
>   Output[s1, 0] := 1: Output[s1, 1] := 1:
>   Output[s2, 0] := 0: Output[s2, 1] := 0:
>   Output[s3, 0] := 0: Output[s3, 1] := 0:
>   Output[s4, 0] := 0: Output[s4, 1] := 0:
>   Start := s0:
>   DFA_ex4 := [ States, InputAlphabet,
>       OutputAlphabet, Trans, Output, Start];
        DFA_ex4 := [{s0, s1, s2, s4, s3}, {0, 1}, {0, 1}, Trans, Output, s0]
```

We have provided the Maple routine DFSMoutput to simulate a finite-state machine with output. We will not discuss the implementation of this routine until after we discuss finite-state machines with no output. However, you can use the routine now without knowing its implementation. DFSMoutput takes three arguments, a finite-state machine with output, an input list, and a boolean for whether or not to print the states of the simulation. The input list is represented by a Maple list. The resulting output list is returned.

Although the formal definition of a finite-state machine requires the transition and output functions to be defined on all possible (state, input symbol) pairs, this is not required in our implementation. The reason is that it can be tedious to type in the entire finite-state machine. If a (state, input symbol) pair is seen during the simulation which is not in the domain of the transition function, the warning message "Not all of the input was seen!" will be printed to the terminal. The simulation will also stop at that point. The portion of the output list that was generated will still be returned. When the output function is not defined for a (state, input symbol) pair, we simply choose to add nothing to the output list.

We will demonstrate the routine on Example 4 on page 667. Having already defined DFA_ex4 above, we would type the following into Maple to obtain the output shown below.

```
> Result := DFSMoutput(DFA_ex4, [1, 0, 1, 0, 1, 1]);

Error, (in DFSMoutput) DFSMoutput
 uses a 3rd argument, psteps, which is missing

> print(Result);
```

Result

The DFSMoutput routine performs some checks on the first argument. These checks are designed to find some common mistakes in representing a finite-state machine with output. These checks include:

- Is the Maple object purporting to represent a finite-state machine a Maple list?
- Does this list consist of 6 components?
- Is the first component of the list a Maple set?
- Is the second component of the list a Maple set?
- Is the third component of the list a Maple set?
- Is the fourth component of the list a Maple table?
- Is the fifth component of the list a Maple table?
- Is the sixth component a member of the first component, where the sixth component represents the starting state and the first component represents the set of states?
- For each object in the domain of the transition function:
  - Is the object a list of 2 components?
  - Is the first component a member of the set of states?

- Is the second component a member of the input alphabet?
- Does the object map to a member of the set of states via the transition function?

- For each object in the domain of the output function:
  - Is the object a list of 2 components?
  - Is the first component a member of the set of states?
  - Is the second component a member of the input alphabet?
  - Does the object map to a member of the output alphabet via the output function?

These checks are encoded in the routine `IsDFSMoutput` which is provided in the ftp archive site. The `DFSMoutput` routine calls `IsDFSMoutput` to perform the check.

10.4 Finite-State Machines with No Output

Finite-state machines can also be used to recognize a language without generating output. Such machines use a set of final states to recognize a string as being part of the language. The finite-state machine is said to recognize the input string if the state the machine is in when all input has been read is a member of the set of final states. Such finite-state machines are also known as finite-state automata.

A finite-state automaton is defined by a set of states, an input alphabet, a transition function, a starting state, and a set of final states. Unlike a finite-state machine with output, these machines do not have an output function or alphabet, but they do have the set of final states. Our Maple representation of a finite-state automaton is similar to our representation of finite-state machines with output, were we use a Maple set for final states. To summarize, a finite-state automaton $M = (S, I, f, s_0, F)$ is represented in Maple as a list of five items:

Item	Type	Description
1	Set	The set of states, S, for M
2	Set	The set of input alphabet symbols, I
3	Table	The transition function, f
4	any type	The starting symbol, s_0
5	Set	The set of final states, F

We now give an example of how to express a particular DFA using Maple, given a figure that defines this machine. Consider the DFA shown in

Figure 10.1. Entries in a Maple table *Delta* are defined by placing the domain elements in square brackets and assigning the range element.

```
>  States := { s0, s1, s2};
```

$$States := \{s0, s1, s2\}$$

```
>  Start := s0;
```

$$Start := s0$$

```
>  Delta := table();
```

$$\Delta := \text{table}([$$
$$])$$

```
>  Delta[s0, a] := s0;
```

$$\Delta_{s0,a} := s0$$

```
>  Delta[s0, b] := s1;
```

$$\Delta_{s0,b} := s1$$

```
>  Delta[s1, c] := s2;
```

$$\Delta_{s1,c} := s2$$

```
>  Delta[s2, a] := s2;
```

$$\Delta_{s2,a} := s2$$

```
>  Delta[s2, b] := s2;
```

$$\Delta_{s2,b} := s2$$

```
>  Delta[s2, c] := s2;
```

$$\Delta_{s2,c} := s2$$

```
>  Alphabet := { a, c, b};
```

$$Alphabet := \{a, c, b\}$$

```
>  FinalStates := { s2 };
```

$$FinalStates := \{s2\}$$

```
>  DFA := [ States, Alphabet, Delta, Start, FinalStates ];
```

$$DFA := [\{s0, s1, s2\}, \{a, c, b\}, \Delta, s0, \{s2\}]$$

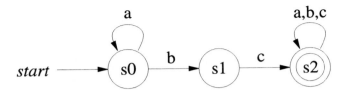

Figure 10.1: An Example DFA

Now that we have a Maple representation for a DFA, we need a routine similar to IsDFSMoutput to check for common errors in the representation. The routine IsDFArecog provides this functionality. The checks it performs are:

- Is the Maple object purporting to represent a finite-state machine a Maple list?
- Does this list consist of 5 components?
- Is the first component of the list a Maple set?
- Is the second component of the list a Maple set?
- Is the third component of the list a Maple table?
- Is the fifth component of the list a Maple set? Is it a subset of the first component, the set of all states?
- Is the fourth component a member of the first component, where the fourth component represents the starting state and the first component represents the set of states?
- For each object in the domain of the transition function:
  - Is the object a list of 2 components?
  - Is the first component a member of the set of states?
  - Is the second component a member of the input alphabet?
  - Does the object map to a member of the set of states via the transition function?

The source for IsDFArecog is provided in the ftp archive.

The function DFArecog is the routine used to recognize a language with a finite automaton. It takes two arguments. The first is the DFA represented as described previously. The second argument is a Maple list for the input. The return value is a Boolean value which tells us whether or not the input was recognized. As was the case with finite-state machines with output, we do not need to define the entire transition function on all (state, input symbol) pairs. If, during the simulation of the DFA, an undefined (state, input symbol) pair is encountered, DFArecog returns the value false.

We will now finish of this section with an example of DFArecog. For the DFA in Figure 10.1, we check for acceptance of the strings listed in the table below. Following the table is output from Maple, where the ":" symbol separates input which has been seen (on the left) from input which has yet to be seen (on the right).

Input No.	Input	Result
1	a	rejected
2	a, b	rejected
3	a, b, b	rejected
4	a, a, b, c, a, a, b, c, b, a, c	accepted
5	b, c	accepted

The example in question is

```
> States    := { s0, s1, s2 };
```
$$States := \{s0, s1, s2\}$$

```
> Alphabet := { a, b, c };
```
$$Alphabet := \{a, c, b\}$$

```
> Delta := table();
```
$$\Delta := \text{table}([$$
$$])$$

```
> Delta[s0, a]  := s0: Delta[s0, b]  := s1:
> Delta[s1, c]  := s2: Delta[s2, a]  := s2:
> Delta[s2, b]  := s2: Delta[s2, c]  := s2;
```
$$\Delta_{s2,c} := s2$$

```
> DFA1 := [ States, Alphabet, Delta, s0, { s2 } ];
```
$$DFA1 := [\{s0, s1, s2\}, \{a, c, b\}, \Delta, s0, \{s2\}]$$

Maple Output for input list 1. We end up on state s0, which is not an accepting state.

```
> DFArecog(DFA1, [a], true);
```
$$Starting\ state, s0$$
$$Next\ state, s0$$
$$a, :$$
$$false$$

Maple Output for input list 2. We end up in state s1, which is not an accepting state.

```
> DFArecog(DFA1, [a, b], true);
```

$$Starting\ state, s0$$
$$Next\ state, s0$$
$$a, :, b$$
$$Next\ state, s1$$
$$a, b, :$$
$$false$$

Maple Output for input list 3. There is no transition on (s1, b) and so we reject the string.

```
> DFArecog(DFA1, [a, b, b], true);
```

$$Starting\ state, s0$$
$$Next\ state, s0$$
$$a, :, b, b$$
$$Next\ state, s1$$
$$a, b, :, b$$
$$false$$

Maple Output for input list 4. The string is accepted in state 2.

```
> DFArecog(DFA1, [a, a, b, c, a, a, b, c, b, a, c], true);
```

$$Starting\ state, s0$$
$$Next\ state, s0$$
$$a, :, a, b, c, a, a, b, c, b, a, c$$
$$Next\ state, s0$$
$$a, a, :, b, c, a, a, b, c, b, a, c$$
$$Next\ state, s1$$
$$a, a, b, :, c, a, a, b, c, b, a, c$$
$$Next\ state, s2$$
$$a, a, b, c, :, a, a, b, c, b, a, c$$
$$Next\ state, s2$$
$$a, a, b, c, a, :, a, b, c, b, a, c$$
$$Next\ state, s2$$
$$a, a, b, c, a, a, :, b, c, b, a, c$$
$$Next\ state, s2$$
$$a, a, b, c, a, a, b, :, c, b, a, c$$
$$Next\ state, s2$$
$$a, a, b, c, a, a, b, c, :, b, a, c$$
$$Next\ state, s2$$
$$a, a, b, c, a, a, b, c, b, :, a, c$$
$$Next\ state, s2$$
$$a, a, b, c, a, a, b, c, b, a, :, c$$
$$Next\ state, s2$$
$$a, a, b, c, a, a, b, c, b, a, c, :$$
$$true$$

Maple Output for input list 5. The input is accepted in s2.

```
> DFArecog(DFA1, [b, c], true);
```

$$Starting\ state, s0$$
$$Next\ state, s1$$
$$b, :, c$$
$$Next\ state, s2$$
$$b, c, :$$
$$true$$

10.5 Deterministic Finite-State Machine Simulation

Until now, we have avoided a discussion about how the simulation of a finite-state machine is implemented. In fact, neither DFSMoutput nor DFArecog discussed so far do any of the simulation themselves. They are "interface" procedures to DFSMengine. DFSMengine is the Maple procedure which performs the simulation for finite-state machines both with and without output. We will first describe how to use DFSMengine, and then discuss its implementation.

DFSMengine takes six arguments and returns a list with three elements. No checks are made on the validity of the arguments. All checks are assumed to be performed by "interface" functions such as DFSMoutput and DFArecog. The six arguments, in order in which they are presented to DFSMengine, are:

1. *start:* The starting state of the finite-state machine.
2. *trans:* The transition function of the finite-state machine, defined on (state, input symbol) pairs. The range of the function is the set of states.
3. *outtab:* The output function of the finite-state machine. The output function works as described in the DFSMoutput section. A finite automata, which recognizes a language and generates no output, would provide an empty function for this argument.
4. *final:* The set fo final states. A finite-state machine with output would provide an empty set for this argument.
5. *input:* A list for the input sequence.
6. *printsteps:* A boolean for whether or not to print out information on the steps of the simulation.

The returned value of DFSMengine is a list with three elements. The elements of the list are:

1. A boolean on whether or not the input was recognized. The input is recognized provided all of the input is used (see (2) below) and that the ending state is in the set of final states.

2. A boolean on whether or not all of the input was used. The simulation returns without using all of the input if a (state, input symbol) pair is observed and is not defined by the transition function.

3. The output list resulting from the simulation. This is the value which is returned by DFSMoutput.

We will now discuss the method of simulating the deterministic finite-state machine. The process begins in the starting state, having seen no input, and with an empty output list. Then, look at the first input symbol, and check the transition function for the next state to enter. If transition is defined, see whether an element should be added to the output list. Now, enter this next state, and look at the second input symbol to find the third state to enter. This process continues either until the transition function is undefined on a (state, input symbol), or until no more input remains. If we have seen all of the input, we must determine whether the last state is a final state. We will then have all of the information from the simulation needed for both an automaton and a finite-state machine with outout.

Notice that if we are in state S_i looking at input symbol a_j in this process, we must check that the transition function is defined on the pair $S_i \times a_j$. In other words, we must check that $S_i \times a_j \in$ domain(transition function). If the transition function is not defined on this pair, we reject the input string. You may recall that we can easily determine the domain of a table with the `indices` function. This method is used in DFSMengine.

```
> DFSMengine := proc(start, trans, outtab, final,
>                    input, printsteps)
>     local transDomain,         # domain of transition func
>           i,                   # index the input for next item
>           nextState, curState, # current and next states
>                                # the DFA machine
>           result,              # resulting list
>           outtabDomain;        # domain of output table
>
>     # initial result is empty
>     result := [];
>     # the domain of the transition function
>     transDomain := { indices(trans) };
>     # assign the domain of the output table to outtabDomain
>     outtabDomain := { indices(outtab) };
>     # the starting state
>     curState := start;
>     if printsteps then
```

```
>            print('Starting state', curState);
>      fi;
>      # proceed over the input, simulating the machine
>      for i from 1 by 1 to nops(input) do
>        if not member( [ curState, op(i, input) ],
>        transDomain ) then
>              RETURN([false, false, result]);
>        fi;
>        # add an element to the result list?
>        if member([curState, op(i, input)], outtabDomain) then
>              result := [ op(result),
>                  outtab[curState, op(i, input)] ];
>        fi;
>        # compute the next state we will be in
>        nextState := trans[ curState, op(i, input) ];
>        # print information about this step ?
>        if printsteps then
>          print('Next state', nextState);
>          print(op(1..i,input),':',op(i+1..nops(input),input));
>          if nops(result) > 0 then
>            print('Result so far:', result);
>          fi;
>        fi;
>        curState := nextState;
>      od;
>      # All input has been seen, and we may or may not be
>      # in a final state
>      if member(curState, final) then
>            RETURN([true, true, result]);
>      else
>            RETURN([false, true, result]);
>      fi;
>   end:
```

10.6 Nondeterministic Finite Automata

Nondeterministic finite-state automata (NFA) differ from their deterministic cousins in that the transition function can map each (state, alphabet symbol) pair to a set of states instead of a single state. Although not mentioned in the text, NFA's commonly differ from DFA's in another important aspect: there may be "free" transitions allowed between states, i.e. transitions which do not use the next input symbol. Such transitions are commonly called ϵ-transitions. Although they are convenient for some applications, ϵ-transitions do not add to the "power" of NFA's, since any NFA with ϵ-transitions can be converted into an NFA without them, and (trivially) vice-versa. Although NFAs may or may not have ϵ-transitions, DFAs never have them. Figure 10.2 shows two NFA's, one with ϵ-transitions and one without, which accept the same language, namely

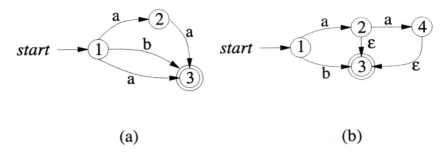

(a)   (b)

Figure 10.2: Figure : Two NFAs accepting the language $\{aa, a, b\}$. Part (b) contains ϵ-transitions.

$\{aa, a, b\}$.

Representing an NFA in Maple is very similar to representing a DFA in Maple, but with two exceptions. First, the elements of the range consists of sets of states, and second, we may have ϵ-transitions. It is in representing ϵ-transitions that the power of having different numbers of arguments for elements in the domain of a table is useful. We may represent an ϵ-transition from state S_i to state S_j simply as `Delta[Si] := { Sj }` . Figure 10.3 shows the NFA represented in Maple below.

```
> Start := s0;
```
$$Start := s0$$

```
> AllStates := {s0,s1,s2,s3,s4,s5,s6,s7,s8,s9,s10};
```
$$AllStates := \{s0, s1, s2, s4, s3, s10, s9, s6, s8, s7, s5\}$$

```
> Accept := {s8,s9,s10}:
> Alphabet := { a, b, c, d }:
> Delta[s0, a] := { s0, s1 }:
> Delta[s0   ] := { s2 }:        # an epsilon transition
> Delta[s1   ] := { s3 }: Delta[s2, a] := { s4 }:
> Delta[s3, b] := { s5 }: Delta[s4, d] := { s5 }:
> Delta[s4   ] := { s7 }: Delta[s5, c] := { s6 }:
> Delta[s6, b] := { s8 }: Delta[s6, d] := { s9 };
```
$$\Delta_{s6,d} := \{s9\}$$

```
> Delta[s7   ] := { s10 }:
> NFA := [ AllStates, Alphabet, Delta, Start, Accept ];
```
$$NFA :=$$
$$[\{s0, s1, s2, s4, s3, s10, s9, s6, s8, s7, s5\}, \{a, c, b, d\}, \Delta, s0, \{s10, s9, s8\}$$
$$]$$

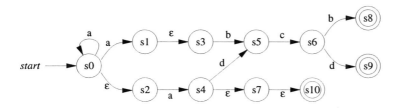

Figure 10.3: An example NFA

Simulating an NFA is slightly more involved than simulating a DFA. When simulating a DFA, the machine can only be in one state during any step of the simulation on a string in the language accepted by the DFA. For NFA's, we must maintain the *set of states* which we might be in, having seen a prefix of a string in the language. To see why, consider the NFA in Figure 10.2a. On input "a", we could be either in state 2 or state 3, or for the NFA in Figure 10.2b, we could also be in states 2 or 3 (we can reach state 3 "for free" via the ϵ-transition between states 2 and 3). Therefore, at any given step in the simulation of an NFA, if we could be in any state in set T_i, we determine the next set of states we could be in on input a, namely T_{i+1}, via the following steps steps:

1. Initially, let $T_{i+1} := \emptyset$
2. For each state $S_k \in T_i$, and on input symbol a, let $T_{i+1} = T_{i+1} \cup f(S_k, a)$, where f is the transition function.
3. Now add to T_{i+1} all states which we can reach from T_{i+1} for "free", i.e. via an ϵ-transition.

Step (2) is typically called a `move` operation, while step (3) is typically called an ϵ-closure operation. A single step in the simulation of an NFA is computed by the composite operation ϵ-closure(move(T_i, a)). To determine whether we accept or reject the string, having seen all of the input, we must determine whether or not the final set of states we could be in contains an accepting state. In other words, we reject the input string if and only if $T_{final} \cap$ final states $= \emptyset$.

We can compute the `move` operation as shown below.

```
>    # NFAmove - A procedure to compute move(T, a)
>    #     NFA    is an NFA
>    #     T      input set of states for the move operation
>    #     a      symbol on which to perform the move
>    #
>    NFAmove := proc(NFA, T, a)
>        local transDomain,
>              i,
```

```
>             result,
>             trans;
>        trans := op(3, NFA);
>        transDomain := { indices(trans) };
>        result := { };
>        for i from 1 by 1 to nops(T) do
>             if member( [op(i, T), a], transDomain ) then
>                 result := result union trans[op(i, T), a];
>             fi;
>        od;
>        RETURN(result);
>   end:
```

The ϵ-closure operation requires a little more work, since we may have many ϵ-transitions in a row. For example, the ϵ-closure of $\{s4\}$ in Figure 10.3 includes states $s7$ and $s10$. Notice that $T \subseteq \epsilon$-closure(T), i.e. every element in the set supplied to the ϵ-closure operation is part of the result.

To compute the ϵ-closure of a set of states, we turn to our old friend, the stack. The idea is to use a stack to hold the states which we have to check for further ϵ-transitions. We copy the top element of the stack, pop it, and then determine which states can be reached from it via ϵ-transitions. All of these added elements must be pushed onto the stack so that they too can be checked for ϵ-transitions. When we have no more elements on the stack, we are done checking states for ϵ-transitions.

Initially, every state in the input state list must be added to the stack. We can do this in one step by converting the set of states directly into a Maple list which will represent the stack. We use the built-in convert function, which converts Maple's built-in data structures from one type to another. Also, we can add multiple elements to the top of the stack in one step by converting the stack and the set to Maple expression sequences, concatenating the sequences, and creating a list from the result as shown on page 82 of the Maple Tutorial text. The resulting Maple procedure for the ϵ-closure operation is as follows:

```
>   # NFAepsClosure - Compute the epsilon procedure of a set
>   # of states in an NFA.  An epsilon transition is identified,
>   # for a delta function delta, as delta[s] for some state s.
>   #     NFA     is an NFA
>   #     T       the set of states needed compute the epsClosure
>   #
>
>   NFAepsClosure := proc(NFA, T)
>     local transDomain,  # domain of transition function
>           toAdd,        # elements to be added to epsClosure
>           epsClosure,   # the epsilon closure
>           stack,        # stack determining when we're done
>           top,          # used to hold the top of the stack
>           trans;
```

```
>    trans := op(3, NFA);
>    transDomain := { indices(trans) };
>    epsClosure := T;
>    stack := convert(T, list);
>    while nops(stack) <> 0 do
>        top := StackTop(stack);
>        stack := StackPop(stack);
>        if member( [top], transDomain ) then
>            toAdd := trans[top] minus epsClosure;
>            stack := [op(toAdd), op(stack)];
>            epsClosure := epsClosure union toAdd;
>        fi;
>    od;
>    RETURN(epsClosure);
> end:
```

Given the move and ϵ-closure procedures, the procedure to accept or reject an input string is similar to that for DFA's and is given below.

```
>    # NFAaccept - accept or reject a string given an NFA.
>    # Return boolean on acceptance.
>    #    NFA is the NFA
>    #    input is the input sequence
>    #    printsteps: do we print simulations ?
>    #
>
>    NFAaccept := proc(NFA, input, printsteps)
>        local transDomain,        # domain (transition func)
>              i,                   # index for next item
>              S,
>              trans;
>
>        # a quick check that NFA is an NFA
>        if not IsNFA(NFA) then
>            print('Not NFA');
>            RETURN(false);
>        fi;
>        # assign some variables for convenience
>        trans := op(3, NFA);
>        transDomain := { indices(trans) };
>        # compute the starting state
>        S := NFAepsClosure(NFA, {op(4, NFA)});
>        if printsteps then
>            print('Starting State Set', S);
>        fi;
>        # now simulate the machine
>        for i from 1 by 1 to nops(input) do
>            S := NFAepsClosure(NFA,NFAmove(NFA,S,op(i,input)));
>            if printsteps then
>                print('Set of states', S);
>                print(op(1..i, input),':',
>                      p(i+1..nops(input), input));
>            fi;
>        od;
```

```
>        if S intersect op(5, NFA) <> {} then
>            RETURN(true);
>        else
>            RETURN(false);
>        fi;
>   end:
```

As with a DFA, we supply a routine which helps determine whether or not a Maple object conforms to our representation of a NFA. This routine is available from the ftp archive.

We finish off this section with a couple of examples. The NFA for Example 7 on page 677 of the text has been coded in Maple below. We then provide a few sample input strings and demonstrate which are accepted and which are rejected.

Input No.	Input	Result
1		accepted
2	1	rejected
3	0, 0, 0	accepted
4	0, 1, 0	rejected
5	0, 0, 1	accepted
6	1, 1	accepted

This is constructed as:

```
> States := { s0, s1, s2, s3, s4 }:
> Alphabet := { 0, 1 }:
> Delta := table():
> Delta[s0, 0] := { s0, s2 }: Delta[s0, 1] := { s1 }:
> Delta[s1, 0] := { s3 }: Delta[s1, 1] := { s4 }:
> Delta[s2, 1] := { s4 }: Delta[s3, 0] := { s3 }:
> Delta[s4, 0] := { s3 }: Delta[s4, 1] := { s3 }:
> NFA1 := [ States, Alphabet, Delta, s0, { s0, s4 } ];
```

$$NFA1 := [\{s0, s1, s2, s4, s3\}, \{0, 1\}, \Delta, s0, \{s0, s4\}]$$

Maple output for input list 1. The input is accepted by s0.

```
> NFAaccept(NFA1, [ ], true);
```

$$Starting\ State\ Set, \{s0\}$$
$$true$$

Maple output for input list 2. Rejected since $\{s1\}cap\{s0, s4\} = \emptyset$.

```
> NFAaccept(NFA1, [1], true);
```

$$Starting\ State\ Set, \{s0\}$$
$$Set\ of\ states, \{s1\}$$
$$1, :$$
$$false$$

Maple output for input list 3. Accepted since $\{s0, s2\}cap\{s0, s4\} = \{s0\} \neq \emptyset$.

> NFAaccept(NFA1, [0, 0, 0], true);

$$Starting\ State\ Set, \{s0\}$$
$$Set\ of\ states, \{s0, s2\}$$
$$0, :, 0, 0$$
$$Set\ of\ states, \{s0, s2\}$$
$$0, 0, :, 0$$
$$Set\ of\ states, \{s0, s2\}$$
$$0, 0, 0, :$$
$$true$$

Maple output for input list 4. Rejected since $\{s3\}cap\{s0, s4\} = \emptyset$.

> NFAaccept(NFA1, [0, 1, 0], true);

$$Starting\ State\ Set, \{s0\}$$
$$Set\ of\ states, \{s0, s2\}$$
$$0, :, 1, 0$$
$$Set\ of\ states, \{s1, s4\}$$
$$0, 1, :, 0$$
$$Set\ of\ states, \{s3\}$$
$$0, 1, 0, :$$
$$false$$

Maple output for input list 5. Accepted since $\{s1, s4\}cap\{s0, s4\} = \{s4\} \neq \emptyset$.

> NFAaccept(NFA1, [0, 0, 1], true);

$$Starting\ State\ Set, \{s0\}$$
$$Set\ of\ states, \{s0, s2\}$$
$$0, :, 0, 1$$
$$Set\ of\ states, \{s0, s2\}$$
$$0, 0, :, 1$$
$$Set\ of\ states, \{s1, s4\}$$
$$0, 0, 1, :$$
$$true$$

Maple output for input list 6. Accepted since $\{s4\}cap\{s0, s4\} = \{s4\} \neq \emptyset$.

> NFAaccept(NFA1, [1, 1], true);

$$Starting\ State\ Set, \{s0\}$$
$$Set\ of\ states, \{s1\}$$
$$1, :, 1$$
$$Set\ of\ states, \{s4\}$$
$$1, 1, :$$
$$true$$

For another example, our NFA is the one shown in Figure 10.3, which recognizes the regular set $(a^*a(b \cup d)c(b \cup d)) \cup (a^*a)$. We present a few example input strings, and the reader is encouraged to try others, while observing how the move and ϵ-closure operations produce the sequence of sets of states which we could be in at each stage as we read the input.

Input No.	Input	Result
1	b	rejected
2	a, a	accepted
3	a, a, b, c, d	accepted

Maple output for input list 1. Rejected since $\{\} \cap \{s8, s9, s10\} = \{\} = \emptyset$.

```
>  NFAaccept(NFA2, [b], true);
```

Not NFA, wrong type for in input argument
Not NFA
false

Maple output for input list 2. Accepted since $\{s1, s3, s2, s4, s7, s10, s0\}$ $\cap \{s8, s9, s10\} = \{s10\} \neq \emptyset$.

```
>  NFAaccept(NFA2, [a, a], true);
```

Not NFA, wrong type for in input argument
Not NFA
false

Maple output for input list 3. Accepted since $\{s9\} \cap \{s8, s9, s10\} = \{s9\} \neq \emptyset$.

```
>  NFAaccept(NFA2, [a, a, b, c, d], true);
```

Not NFA, wrong type for in input argument
Not NFA
false

10.7 Constructing Deterministic Finite Automata Equivalent to a Nondeterministic Finite Automata

NFAs are as powerful as DFAs, which is to say that there is no language accepted by an NFA which cannot be accepted by a DFA. Thus, given any NFA, we can find a DFA which accepts the same language. In this section, we present an algorithm which converts any NFA into an equivalent DFA. In the section on nondeterministic finite automata, we saw how to simulate the action of an NFA on an input string. On each input symbol, we moved from one set of states to another set of states. If we consider each set of states as a new state, we could then view the simulation of an NFA on the original sets of states as a simulation of a DFA on the new states.

Informally, the algorithm works as follows. The start state of the DFA represents all of the states of the NFA which can be reached from the start state of the NFA, using ϵ-transitions alone. This is accomplished with an ϵ-closure operation on the set containing only the starting state of the NFA, i.e.

$$DFAStartState := \epsilon - closure(\{NFAStateState\}) \qquad (10.1)$$

Once we have the starting state of the DFA, we build the transition function and the set of states of the DFA by finding new states and expanding them until all states of the DFA have been expanded. The expansion of a DFA state, S, involves computing ϵ-closure(move(S, i)), for each symbol i in the input alphabet. During each such step, the entries in the transition function are defined as $f(S, i) = \epsilon$-closure(move(S, i)). We keep track of which states have to be expanded by placing newly-discovered states on a stack. This leads us to the following function for converting an NFA into a DFA:

```
>   # NFA2DFA - convert an NFA into a DFA
>   #      The input argument is an NFA, and a DFA is returned
>   #      which accepts the same language
>   #
>
>   NFA2DFA := proc(NFA)
>       local DFAStates,        # all DFA states
>             DFAStart,         # DFA starting state
>             DFArecog,         # accepting states for the DFA
>             DFATrans,         # transition function for the DFA
>             MaybeNewState,    # candidate DFA state
>             i,
>             Alphabet,         # the underlying alphabet
>             top,
>             letter,
>             Stack;
>       # check whether or not the argument really is an NFA
>       if not IsNFA(NFA) then
>           ERROR('Not NFA when converting from NFA to DFA');
>       fi;
>       # the alphabet for the DFA is the same as the NFA
>       Alphabet := op(2, NFA);
>
>       # Compute DFAstart, initialize DFAStates and DFATrans
>       DFAStart := NFAepsClosure(NFA, { op(4, NFA) });
```

```
>         DFAStates := { DFAStart };
>         Stack := StackPush(Stack, DFAStart);
>         DFATrans := table();
>
>         # now, build the DFA
>         while nops(Stack) <> 0 do
>           top := StackTop(Stack);
>           Stack := StackPop(Stack);
>           for i from 1 by 1 to nops(Alphabet) do
>             letter := op(i, Alphabet);
>             MaybeNewState :=
>                 NFAepsClosure(NFA, NFAmove(NFA, top, letter));
>             if nops(MaybeNewState) > 0 and
>                 not member(MaybeNewState, DFAStates) then
>                 Stack := StackPush(Stack, MaybeNewState);
>                 DFAStates := DFAStates union { MaybeNewState };
>                 DFATrans[top, letter] := MaybeNewState;
>             fi;
>           od;
>         od;
>
>         # finally, look dor accepting states.  Any DFA
>         # state which is an NFA accepting state, is an
>         # accepting state of the DFA
>         DFArecog := {};
>         for i from 1 by 1 to nops(DFAStates) do
>           if nops(op(i,DFAStates) intersect op(5,NFA))>0 then
>                 DFArecog := DFArecog union {op(i,DFAStates)};
>                 print('DFArecog now', DFArecog);
>           fi;
>         od;
>         RETURN( [DFAStates,Alphabet,
>             eval(DFATrans),DFAStart,DFArecog]);
>   end:
```

We now finish off this section with a couple of example conversions.

Example We convert from the NFA of Example 7 of the text on page 677 to an equivalent DFA . (See Example 9 on page 679 of the text.)

```
>   DFA := NFA2DFA(NFA1);
```

$$DFA\,Accept\ now, \{\{s4\}\}$$
$$DFA\,Accept\ now, \{\{s4\}, \{s0, s2\}\}$$
$$DFA\,Accept\ now, \{\{s4\}, \{s0, s2\}, \{s0\}\}$$
$$DFA\,Accept\ now, \{\{s4\}, \{s0, s2\}, \{s0\}, \{s1, s4\}\}$$
$$DFA\,Accept\ now, \{\{s4\}, \{s0, s2\}, \{s0\}, \{s1, s4\}, \{s4, s3\}\}$$
$$DFA := [\{\{s3\}, \{s4\}, \{s0, s2\}, \{s1\}, \{s0\}, \{s1, s4\}, \{s4, s3\}\}, \{0, 1\},$$

$$table([$$

$$(\{s4\}, 1) = \{s3\}$$
$$(\{s1, s4\}, 1) = \{s4, s3\}$$
$$(\{s0, s2\}, 0) = \{s0, s2\}$$
$$(\{s4\}, 0) = \{s3\}$$

$(\{s4, s3\}, 0) = \{s3\}$

$(\{s1, s4\}, 0) = \{s3\}$

$(\{s1\}, 1) = \{s4\}$

$(\{s0\}, 1) = \{s1\}$

$(\{s1\}, 0) = \{s3\}$

$(\{s0\}, 0) = \{s0, s2\}$

$(\{s0, s2\}, 1) = \{s1, s4\}$

$(\{s4, s3\}, 1) = \{s3\}$

$(\{s3\}, 0) = \{s3\}$

$]), \{s0\}, \{\{s4\}, \{s0, s2\}, \{s0\}, \{s1, s4\}, \{s4, s3\}\}]$

Example For the NFA in Figure 10.3. To build this example, use the commands

```
> Start := s0:
> AllStates := { s0, s1, s2, s3, s4, s5, s6, s7, s8, s9, s10 };
        AllStates := {s0, s1, s2, s4, s3, s10, s9, s6, s8, s7, s5}

> Accept := { s8, s9, s10 }:
> Alphabet := { a, b, c, d}:
> Delta := table():
> Delta[s0, a] := { s0, s1 }:
> Delta[s0   ] := { s2 }:              # an epsilon transition
> Delta[s1   ] := { s3 }:
> Delta[s2, a] := { s4 }: Delta[s3, b] := { s5 }:
> Delta[s4, d] := { s5 }: Delta[s4   ] := { s7 }:
> Delta[s5, c] := { s6 }: Delta[s6, b] := { s8 }:
> Delta[s6, d] := { s9 }: Delta[s7   ] := { s10 }:
> NFA2 := [ AllStates, Alphabet, Delta, Start, Accept ];
NFA2 :=
```

$[\{s0, s1, s2, s4, s3, s10, s9, s6, s8, s7, s5\}, \{a, c, b, d\}, \Delta, s0, \{s10, s9, s8\}$
$]$

```
> DFA := NFA2DFA(NFA2);
```

$DFA\,Accept\ now, \{\{s8\}\}$

$DFA\,Accept\ now, \{\{s8\}, \{s9\}\}$

$DFA\,Accept\ now, \{\{s8\}, \{s9\}, \{s0, s1, s2, s4, s3, s10, s7\}\}$

$DFA := [\{\{s5\}, \{s6\}, \{s8\}, \{s9\}, \{s0, s2\}, \%1\}, \{a, c, b, d\}, \mathrm{table}([$

$(\%1, d) = \{s5\}$

$(\{s0, s2\}, a) = \%1$

$(\{s6\}, d) = \{s9\}$

$(\%1, a) = \%1$

$(\{s6\}, b) = \{s8\}$

$(\%1, b) = \{s5\}$

$(\{s5\}, c) = \{s6\}$

$]), \{s0, s2\}, \{\{s8\}, \{s9\}, \%1\}]$

$\%1 := \{s0, s1, s2, s4, s3, s10, s7\}$

10.8   Converting Regular Expressions to/from Finite Automata

Conversion from a regular expression to a finite automata can be accomplished via a method known as Thompson's construction. We will provide an outline of Thompson's construction; a more thorough discussion can be found in [Aho, Sethi, Ullman, Compilers, Principles, Techniques, and Tools, 1985, Addison-Wesley].

Before we can talk about Thompson's construction, we must first describe how we can represent a regular expression using Maple. We do this by using lists, sets, and functions. In the scope of regular expressions, a list of Maple objects, `list`, represents the concatenation of the regular expressions represented by `op(1, list)`, `op(2, list)`, ..., `op(nops(list), list)`. A set of Maple objects correspondingly represents the union of regular expressions represented by the elements of the set. The Kleene closure of a regular expression, represented by a Maple object `x`, is represented by `f(x)`, where `f` is some function not defined on `x`. The following example shows the Maple representation of a regular expression.

Example XXX: Give the representation of $(abc \cup def)(gh)^*$ on alphabet $\{a, b, c, d, e, f, g, h\}$.

```
> [ { [a, b, c], [d, e, f] }, Kleene( [g, h] ) ];
```
$$[\{[a, b, c], [d, e, f]\}, \text{Kleene}([g, h])]$$

We will now describe Thompson's construction. For this discussion, let us assume that f is the function for converting a regular expression into an NFA. The algorithm for the construction is recursive, with subexpressions first being converted into NFA's with f, and then a resulting NFA being constructed, by f, by "gluing" together the sub-NFA's, often with ϵ-transitions. There are four cases to consider, as shown in Figure 10.4. NFA's for subexpressions are shown as an ellipse with a circle at either end of the ellipse. The circle on the left represent the starting state of the sub-NFA, and the circle on the right represents the final state of the sub-NFA. The starting state of the resulting NFA is labeled "s", and the resulting final state is labeled "f". In (a), we build an NFA for a symbol x in the input alphabet with a single edge between two states. In (b), given $R_1 R_2 \cdots R_n$, we first construct NFA's for each R_i as NFA$_i$. We then join the sub-NFA's into a single NFA by adding a new start state and linking the NFA's via ϵ-transitions. Notice that the final state of the resulting NFA is the final state of R_n. In (c), we show the resulting NFA for $R_1 \cup R_2 \cup \cdots \cup R_n$. Again, we first build sub-NFA's and build the resulting NFA with ϵ-transitions. Both the starting state and final state are new. In (d) is the result of Kleene closure. The bottom-most ϵ-transition allows for zero instances of the regular expression.

Our Maple implementation of Thompson's construction uses two functions, DONTUSE_Regex2NFA and Regex2NFA. The user of this invokes Regex2NFA, which initializes some variables, calls DONTUSE_Regex2NFA to construct the transition function and set of states, and then combines the results into an NFA. The input alphabet must consist of strings and integers only.

Most computation for the construction is performed by the recursive function DONTUSE_Regex2NFA, named so as to discourage users from calling it directly. It considers each of the four cases for Thompson's construction to build the transition function and the set of states. The states are integers of increasing value as the construction proceeds. The Maple code is as follows:

```
>   DONTUSE_Regex2NFA := proc(rx, Alphabet, Statebase,
>                             argDelta, argStates, pstps)
>     local i, start, finish, finishSet, tmp, Delta, States;
>
>     Delta  := argDelta;   # to update the values
>     States := argStates;  # to update the values
>
>     # check for a member of the alphabet
>     if type(rx, integer) or type(rx, string) then
>         if pstps then
>             print('Atomic', rx);
>         fi;
>           if not member(rx, Alphabet) then
>               ERROR('Not member of alphabet', rx);
>         fi;
>         start  := Statebase;
>         finish := Statebase+1;
>         Delta[Statebase, rx] := { Statebase+1 };
>         States := States union { Statebase, Statebase+1 };
>         RETURN([Statebase+1, eval(Delta), States]);
>     fi;
>
>     # check for concatenation
>     if type(rx, list) then
>         if pstps then
>             print('Concatenation', rx);
>         fi;
>         finish := Statebase;
>         # build the NFA by concatenating individual NFAs
```

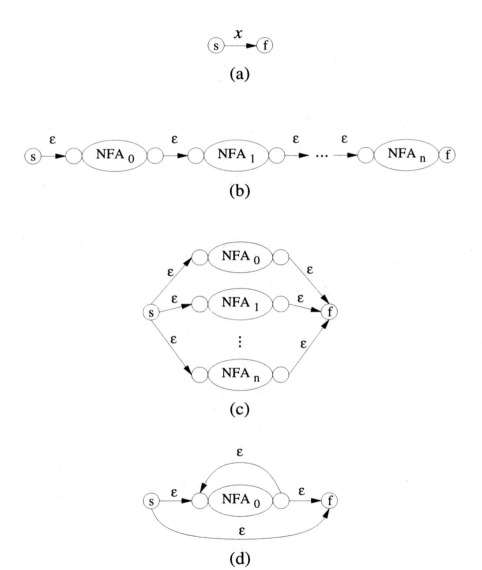

Figure 10.4: Outline of Thompson's construction to build an NFA from a regular expression.

```
>            for i from 1 by 1 to nops(rx) do
>                # convert a component into a regex
>                Delta[finish] := { finish+1 };
>                tmp := DONTUSE_Regex2NFA(op(i, rx), Alphabet,
>                            finish+1, eval(Delta), States, pstps);
>                finish := op(1, tmp);
>                Delta  := op(2, tmp);
>                States := op(3, tmp) union { Statebase };
>            od;
>            # ok, we are done
>            RETURN([finish, eval(Delta), States]);
>        fi;
>
>        # check for disjunction
>        if type(rx, set) then
>            if pstps then
>                print('Disjunction', rx);
>            fi;
>            start := Statebase;
>            finish := Statebase;
>            finishSet := {};
>            Delta[start] := { start+1 };
>            for i from 1 by 1 to nops(rx) do
>                Delta[start] := Delta[start] union { finish+1 };
>                tmp := DONTUSE_Regex2NFA(op(i, rx), Alphabet,
>                        finish+1, eval(Delta), States, pstps);
>                finish := op(1, tmp);
>                Delta  := op(2, tmp);
>                States := op(3, tmp);
>                finishSet := finishSet union { finish };
>            od;
>            finish := finish+1;
>            States := States union { finish };
>            for i from 1 by 1 to nops(finishSet) do
>                Delta[op(i, finishSet)] := { finish };
>            od;
>            RETURN([finish, eval(Delta), States]);
>        fi;
>
>        # check for closure
>        if type(rx, function) then
>            if pstps then
>                print('Closure', rx);
>            fi;
>            start := Statebase;
>            tmp := DONTUSE_Regex2NFA(op(1, rx), Alphabet,
>                        start+1, eval(Delta), States, pstps);
>            finish := op(1, tmp) + 1;
>            Delta  := op(2, tmp);
>            States := op(3, tmp);
>            Delta[start] := { start+1, finish };
>            Delta[finish-1] := { finish, start+1 };
>            States := States union { start, finish };
>            RETURN([finish, eval(Delta), States]);
```

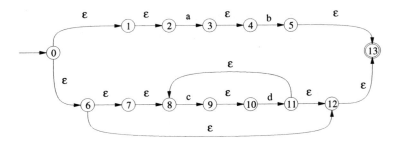

Figure 10.5: NFA generated for $ab \mid (cd)^*$

```
>     fi;
>
>     # if we get here, we have a problem
>     ERROR('ERROR IN REGULAR EXPRESSION REPRESENTATION!');
>   end:
```

The procedure Regex2NFA appears below.

```
>   Regex2NFA := proc(rx, Alphabet, printsteps)
>       local tmp, Delta;
>
>       Delta := table();
>       tmp := DONTUSE_Regex2NFA(rx, Alphabet, 0,
>                    Delta, {0}, printsteps);
>       RETURN([op(3, tmp), Alphabet, op(2, tmp),
>              0, { op(1, tmp) }]);
>   end:
```

We now finish off this section with an example.

10.9 Turing Machines

In this section, we describe how to simulate the actions of Turing machines using Maple. We will only cover deterministic Turing machines (DTMs). We start off by describing how to simulate the actions of a Turing machine with an unbounded tape, and then how to use one with a deterministic Turing machine.

A Turing machine tape consists of storage space and a head. The storage space can be represented in Maple as a list, and the head as an index in the list. Our blank symbol will always be the space character, represented in Maple as ' '. The integer representing the position of the head, along with the list representing the contents of the tape, can themselves be kept in a list. We will provide a Maple routine, IsTmTape, which performs some checks on a Maple object to see whether it represents the tape of a Turing machine conforming to this representation.

Our tape has operations `moveLeft`, `moveRight`, `read`, `new`, and `write`. The move operations will re-position the head. It may be necessary to add a blank character to the beginning or end of the list to perform the move. The `read` and `write` operations are for reading and writing symbols at the current head position. The `new` operation produces an all-blank tape.

```
> IsTmTape := proc(MaybeTmTape)
>     local i;
>
>     if not type(MaybeTmTape, list) then
>         print('Not Tm Tape, must be a list');
>         RETURN(false);
>     fi;
>     if not type(op(1, MaybeTmTape), integer) then
>         print('Not Tm Tape, first opmust be an integer');
>         RETURN(false);
>     fi;
>     if not type(op(2, MaybeTmTape), list) then
>         print('Not Tm Tape, second op must be a list');
>         RETURN(false);
>     fi;
>     i := op(1, MaybeTmTape);
>     if i < 1 or i > nops(op(2, MaybeTmTape)) then
>         print('Not Tm Tape, index is out of range');
>         RETURN(false);
>     fi;
>     RETURN(true);
> end:
```

Constructing a new Turing machine tape amounts to creating a single-element list with the blank symbol, with the index for the head at position 1. The following Maple routine produces a new, blank Turing machine tape.

```
> TmTape_New := proc()
>     RETURN([1, [' ']]);
> end:
```

The move operations append a blank character to a list when the index is 1 during a move left, or is nops(list) during a move right. Remember that we can add a symbol to the end or beginning of a list by first converting it to a Maple expression sequence, adding the blank character, and then creating a new list. The move operations are shown in the following:

```
> TmTape_MoveLeft := proc(tape)
>     if not IsTmTape(tape) then
>         ERROR('Not Tm Tape during move left');
>     fi;
>     if op(1, tape) = 1 then
```

```
>              RETURN([1, [' ', op( op(2, tape) )]]);
>        else
>              RETURN([op(1,tape)-1, op(2, tape)]);
>        fi;
> end:
>
> #
> # TmTape_MoveRight - move head one position to the right
> #
>
> TmTape_MoveRight := proc(tape)
>     if not IsTmTape(tape) then
>         ERROR('Not Tm Tape during move right');
>     fi;
>     if op(1, tape) = nops(op(2, tape)) then
>         RETURN([nops(op(2, tape))+1,[op(op(2,tape)),' ']]);
>     else
>         RETURN([op(1,tape)+1, op(2, tape)]);
>     fi;
> end:
```

Finally, the write operations uses the Maple subsop command to replace an element; we need only perform a simple op command to obtain the element currently at the head position. The routines for reading and writing from a Turing machine tape are as follows:

```
> TmTape_Read := proc(tape)
>     if not IsTmTape(tape) then
>         ERROR('Not Tm Tape during read');
>     fi;
>     RETURN(op( op(1,tape), op(2,tape) ));
> end:
```

This returns the symbol found at the current head position.

```
> TmTape_Write := proc(tape, symbol)
>    if not IsTmTape(tape) then
>        ERROR('Not Tm Tape during write');
>    fi;
>    if nargs <> 2 then
>        ERROR('Wrong number arguments to TmTape_Write');
>    fi;
>    RETURN([op(1,tape),subsop(op(1,tape)=symbol,
>             op(2,tape))]);
> end:
```

This returns the updated tape.

Our representation for a Turing machine, like the representations for automata and Turing machine tapes, is a Maple list. The list for a Turing machine has four components. The first component is the set of states for the Turing machine and the second component is the alphabet. The third component is the transition function, which maps a (state, input symbol) pair onto a (state, input symbol, direction) triple. The final component is the starting state of the Turing machine. There following routine checks its argument for a Turing machine representation.

We provide a routine IsDTM, available from the ftp archive, which tells whether a Maple object conforms to our representation of a Turing machine.

We are now ready to discuss how to "run" a Turing machine. We cannot truly build a Turing machine. Although the tape of a Turing machine is "infinite", computers which humans build have a limit on the amount of memory they can have. Fortunately, we can study Turing machines without requiring an infinite tape by limiting the number of "steps" we may simulate a Turing machine. By having a limit on the number of steps, we are placing a limit on the amount of space a Turing machine can use while being simulated. To see why, remember that a Turing machine can move only one **space** left or right during one step of the machine. In the worst case, we could move always left, or always right, to use as much space as possible. In doing so, we may only read or write from n squares after n turns.

Our function, called DTMRun, will take as arguments a Turing machine, an input tape, a maximum number of steps to simulate the machine, and an argument which allows us to print the steps the Turing machine makes when run. Having a maximum number of steps to simulate the machine is important. This is because we cannot know when a given Turing machine will halt, or even if it will halt, on a given input tape. The process for simulating one step of the Turing machine is:

1. Read the current symbol under the head of the tape.
2. Evaluate the transition function on the current state and the input symbol read during (1). If the transition function is not defined on the current state and input symbol, then halt.
3. The range element obtained by evaluating the transition function is a triple consisting of a new current state, a tape symbol to write to tape, and a direction to move the tape head. From this range element, update the tape by writing the symbol and then moving the tape head either left or right.
4. Also from the range element, update the current state.

The Maple routine DTMRun uses the above process for each step in simulating the Turing machine, while using a "for" loop to insure the maximum number of steps is not exceeded. The return value is a Maple list with three members. The first member is boolean, and is true if the Turing machine halted, and false if we have returned because the maximum number

of steps was reached. The second member is the last "current state" the
machine was in when we returned. The third member is the tape which
resulted by running the machine. We present the Maple code for the function below.

```
> DTMRun := proc(DTM, ArgTape, MaxSteps, printsteps)
>     local transDomain, # domain transition function
>           i,           # index of the input for next item
>           curState,    # current state
>           rangeElem,   # a range element for trans()
>           trans,       # the transition function
>           Tape,        # a local copy of the tape
>           tapeSymbol,  # holds a symbol from the tape
>           tmp;
>     # a quick check that the args are a DTM and a tape
>     if IsDTM(DTM) = false or IsTmTape(ArgTape) = false then
>         ERROR('Arguments are wrong, aborting');
>     fi;
>     if type(MaxSteps, integer) = false then
>      ERROR('The maximum number of steps must be an integer');
>     fi;
>     # assign trans and transDomain, used for convenience
>     Tape := ArgTape;
>     trans := op(3, DTM);
>     transDomain := { indices(trans) };
>     curState := op(4, DTM);    # the starting state
>     # proceed over the input, simulating the machine
>     for i from 1 by 1 to MaxSteps do
>         if printsteps then
>             lprint('Current State', curState);
>             lprint('Tape', Tape);
>         fi;
>         tapeSymbol := TmTape_Read(Tape);
>         if not member([curState,tapeSymbol],transDomain) then
>             RETURN([true, curState, Tape]);
>         fi;
>         rangeElem := [ trans[curState, tapeSymbol] ];
>         Tape := TmTape_Write(Tape, op(2, rangeElem));
>          tmp := op(2, rangeElem);
>         Tape := TmTape_Write(Tape, tmp);
>         if op(3, rangeElem) = LEFT then
>           Tape := TmTape_MoveLeft(Tape);
>         else
>           Tape := TmTape_MoveRight(Tape);
>         fi;
>         curState := op(1, rangeElem);
>     od;
>     # reached the maximum number of steps
>     RETURN([false, curState, Tape]);
> end:
```

We can use the return value from DTMRun for both Turing machines which
compute a function and which accept a language. We will consider both

of these cases, starting with Turing machines which accept a language.

If the first component of the return value of DTMRun is true, then the Turing machine halted and we can either accept or reject the string based on the second component being in a set of final states. If the first component is false, then DTMRun returned because the maximum number of steps was reached, and we can neither reject nor accept the input string. We can try again by increasing the maximum number of steps the Turing machine may run. However, there are Turing machines which will never halt on some or all of its inputs, and so no maximum number of steps will allow the Turing machine to halt.

For Turing machines which compute a function, we must again check the first component of the return value. If it is true, then we can say that the Turing machine computes the resulting tape found in the third component, given the tape that was provided as input to DTMRun. If the first component if false, we again cannot say anything. We are unable to determine whether the function the Turing machine computes is defined on the input value, or that we simply needed to simulate the Turing machine for more steps.

An example of a Turing machine designed to recognize a language is Example 3 of the text. The Turing machine recognizes the set $\{0^n1^n \mid n \geq 1\}$. The accepting state for this machine is state $s6$. The machine, expressed in our Maple representation is shown in the following Maple input:

```
>   States := {s0, s1, s2, s3, s4, s5, s6}:
>   Delta := table():
>   Delta[s0, 0  ] := (s1, M  , RIGHT):
>   Delta[s1, 0  ] := (s1, 0  , RIGHT):
>   Delta[s1, 1  ] := (s1, 1  , RIGHT):
>   Delta[s1, M  ] := (s2, M  , LEFT ):
>   Delta[s1, ' '] := (s2, ' ', LEFT ):
>   Delta[s2, 1  ] := (s3, M  , LEFT ):
>   Delta[s3, 1  ] := (s3, 1  , LEFT ):
>   Delta[s3, 0  ] := (s4, 0  , LEFT ):
>   Delta[s3, M  ] := (s5, M  , RIGHT):
>   Delta[s4, 0  ] := (s4, 0  , LEFT ):
>   Delta[s4, M  ] := (s0, M  , RIGHT):
>   Delta[s5, M  ] := (s6, M  , RIGHT):
>   Alphabet := { 0, 1, M, ' ' }:
>   DTM_ex3 := [States, Alphabet, Delta, s0];
        DTM_ex3 := [{s0, s1, s2, s4, s3, s6, s5}, {0, 1, M, }, \Delta, s0]
```

To run this example, tracing the intermediate steps, use the command

```
>   DTMRun( DTM_ex3 , [1,[0,0,0,1,1,1]], 30 , true );
```

This produces the output It finishes in state $s6$, and shows that the tape contains all "M" symbols as described in the text.

10.10 Computations and Explorations

Exercise 1. The Busy Beaver problem, described in the problems section of the text, involves finding the largest number of 1's which can be written onto a tape given a Turing machine with n states (and a halting state) and alphabet consisting of 1 and B.

Solution: One method of finding solutions to the busy beaver problem is to run all Turing machines with n states. We must run the Turning machines until they either halt or have reached a certain number of steps. The later condition is important, since some of the Turing machines will not halt.

We will present a solution for $n = 2$ states. Our alphabet, starting state, and set of states is known. The only thing which will change for our Turing machines is the transition function. For each (state, input symbol) pair, we must assign a triple (state, output symbol, direction). One solution would be to iterate over all possible triples for each (state, input symbol) pair in a "nested loop" fashion. The possible (state, input symbol) pairs are (s1, ' '), (s1, 1), (s2, ' '), and (s2, 1). The possible triples include (s1, ' ', RIGHT), (s1, ' ', LEFT), (s1, 1, RIGHT), (s1, 1, LEFT), (s2, ' ', RIGHT), (s2, ' ', LEFT), (s2, 1, RIGHT), (s2, 1, LEFT), and (HALT, 1, LEFT). The last triple allow us to enter the halting state, printing a '1', and arbitrarily moving left. Since the Turing machine halts when it enters the halting state, we could just as readily defined the last triple as (HALT, 1, RIGHT).

Each Turing machine we compute as we perform the nested looping is given a blank tape and simulated for a finite number of steps. If the Turing machine halted, we should count the number of 1's left on the output tape. To do this, we use a function `FindNumOfOnes`. Each time we find a Turing machine which halted with more 1's than a previous Turing machine, we output it.

```
>   FindNumOfOnes := proc(tape)
>       local i, i_max, count, contents;
>
>       contents := op(2, tape);
>       i_max := nops(contents);
>       count := 0;
```

```
>        for i from 1 by 1 to i_max do
>            if op(i, contents) = 1 then
>                count := count+1;
>            fi;
>        od;
>        RETURN(count);
>    end:
>
>    FindBeaver2 := proc()
>        local i1, i2, j1, j2, allStates, inputAlphabet,
>            tape, delta, result, dtm, numOfOnes, maxOnes,
>            setOfDestTriples, sizeSetOfDestTriples;
>
>        allStates := { s1, s2, HALT};
>        inputAlphabet := { ' ', 1 };
>        setOfDestTriples := { [s1, ' ', RIGHT],
>            [s1, ' ', LEFT], [s1, 1,   RIGHT],
>            [s1, 1,   LEFT], [s2, ' ', RIGHT],
>            [s2, ' ', LEFT], [s2, 1,   RIGHT],
>            [s2, 1,   LEFT], [HALT, 1, LEFT] };
>        sizeSetOfDestTriples := nops(setOfDestTriples);
>        maxOnes := 0;
>
>        for i1 from 1 by 1 to sizeSetOfDestTriples do
>          for i2 from 1 by 1 to sizeSetOfDestTriples do
>           for j1 from 1 by 1 to sizeSetOfDestTriples do
>            for j2 from 1 by 1 to sizeSetOfDestTriples do
>              delta := table();
>              delta[s1, ' '] := op(op(i1, setOfDestTriples));
>              delta[s1, 1 ] := op(op(i2, setOfDestTriples));
>              delta[s2, ' '] := op(op(j1, setOfDestTriples));
>              delta[s2, 1 ] := op(op(j2, setOfDestTriples));
>              tape := TmTape_New();
>              dtm := [allStates,inputAlphabet,eval(delta),s1];
>              result := DTMRun(dtm, tape, 100, false);
>              if op(1,result) then
>                tape := op(3, result);
>                numOfOnes := FindNumOfOnes(tape);
>                if numOfOnes > maxOnes then
>                  print('Best So Far', numOfOnes);
>                  maxOnes := numOfOnes;
>                  print(dtm);
>                fi;
>              fi;
>            od;
>           od;
>          od;
>        od;
>    end:
```

Exercise 2. Having a computer iterate over all possible Turing machines for $n=2$ is time consuming. Finding the solution to the Busy Beaver problem for $n=3$ requires much, much more time than solving the problem for

$n=2$, and the time grows at a fantastically rapid rate as n grows. This makes finding solutions for even small values of n, such as $n = 6$, by exhaustively checking all Turning machines infeasible.

Solution: Here is a solution for for $n=3$:

```
> # Define the Busy Beaver problem for n=3 non-halting states
> States := {0, 1, 2, HALT};
```
$$States := \{0, 1, 2, HALT\}$$

```
> Delta := table():
> Delta[0, ' '] := (1,    1,   LEFT ):
> Delta[0, 1  ] := (HALT, 1,   LEFT ):
> Delta[1, ' '] := (2,    ' ', LEFT ):
> Delta[1, 1  ] := (1,    1,   LEFT ):
> Delta[2, ' '] := (2,    1,   RIGHT):
> Delta[2, 1  ] := (0,    1,   RIGHT):
> Beaver3 := [States, {' ', 1}, Delta, 0];
```
$$Beaver3 := [\{0, 1, 2, HALT\}, \{1, \ \}, \Delta, 0]$$

```
>
> # Now run the machines for some output.
> Tape := TmTape_New();
```
$$Tape := [1, [\ ]]$$

```
> Result3 := DTMRun(Beaver3, Tape, 500, false);
```
$$Result3 := [true, HALT, [3, [1, 1, 1, 1, 1, 1]]]$$

Exercise 3. 5. We provide a solution for $n = 4$ here. Exhausting check all cases is infeasible without an extremely efficient program.

Solution:

```
> # Define the Busy Beaver problem for n=4 non-halting states
> States := { 0, 1, 2, 3, HALT }:
> Delta := table():
> Delta[0, ' '] := (2,    1,   RIGHT):
> Delta[0, 1  ] := (2,    1,   LEFT ):
> Delta[1, ' '] := (HALT, 1,   LEFT ):
> Delta[1, 1  ] := (3,    1,   LEFT ):
> Delta[2, ' '] := (0,    1,   LEFT ):
> Delta[2, 1  ] := (1,    ' ', LEFT ):
> Delta[3, ' '] := (3,    1,   RIGHT):
> Delta[3, 1  ] := (0,    ' ', RIGHT):
> Beaver4 := [States, {' ', 1}, Delta, 0];
```
$$Beaver4 := [\{0, 1, 2, 3, HALT\}, \{1, \ \}, \Delta, 0]$$

We run this machine as follows:

```
>  Tape := TmTape_New();
```
$$Tape := [1, [\ ]]$$

```
>  Result4 := DTMRun(Beaver4, Tape, 500, false);
```
$$Result4 := [true, HALT, [1, [\ , 1, \ , 1, 1, 1, 1, 1, 1, 1, 1, 1, 1, 1, 1]]]$$

10.11 Exercises/Projects

Exercise 1. Construct a Maple procedure for simulating the action of a Moore machine. (See page 671 of the text for a definition.)

Exercise 2. Develop Maple procedures for finding all the states of a finite-state machine that are reachable from a given state and for finding all transient states and sinks of the machine. (See page 706 of the text for definitions.)

Exercise 3. Construct a Maple procedure that computes the star height of a regular expression. (See page 706 of the text for a definition.)

Exercise 4. Represent the Turing machine for adding two nonnegative integers given in Example 4 on page 699 of the text. Test that this Turing machine produces the desired result for some sample input values.

Exercise 5. Construct a Maple procedure that simulates the action of a Turing machine that may move right, left, or not at all at each step and that may have more than one tape.

Exercise 6. Construct a Maple procedure that simulates the action of a Turing machine with a two-dimensional tape.

Index

&and, 32, 99, 324, 325
&implies, 33
&not, 32, 99, 324, 326
&or, 32, 324
->, 108, 109
closure, 367, 370

addedge, 191, 218–220, 258
addvertex, 218, 219, 284
adjacency, 246
alias, 73
allpairs, 191, 259
allstructs, 118
ancestor, 276
AND2, 31
AND, 27–29, 31
and, 27, 324, 325, 334
appendto, 341
ArityBalanced, 279
array, 176
arrivals, 231

BackColor, 303–306
bequal, 32, 328, 348
BigFib, 174
binomial, 120–122
BinSearch, 163
Bipartite, 238, 239
Birthdays, 144
BiSearch, 285
boolean, 323
bottom up, 110
Breadth, 302
bsimp, 88, 100, 335, 336, 343–345, 348
BubbleSort, 297, 299

canon, 329, 330, 342
ceil, 49
choose, 118, 119, 136
chrem, 66
chrompoly, 263

City1, 314
CNF, 330, 342
coeff, 124, 125
colon, 12–14
combinat[fibonacci], 174
combinat, 19, 20, 39, 50, 117, 118, 123, 128, 131, 136, 174, 190, 265, 340
combstruct, 21, 117–119, 128–130, 134
compatibility, 21
complement, 242
complete, 232, 237, 238
components, 251, 270, 275, 313
Concentric, 222, 223
connectivity, 267
connect, 218
contract, 228
convert, 83, 122, 348
cost, 60, 61
count, 118, 173
CryptChar, 77
cube, 234
cyclebase, 275

daughter, 218, 277
dcmin, 337, 347, 348
DegCount, 230, 231
degree, 37
degreeseq, 249
delete, 227
departures, 231
DFArecog, 358, 359, 361
DFSMengine, 361, 362
DFSMoutput, 354–356, 361
DiConnect, 252
dinic, 263, 264
DirectedDegCount, 231
DividesRelation, 184, 205
divisors, 205
djspantree, 218
DNF, 330, 342

dodecahedron, 236
draw, 19, 118, 209, 210, 219–222, 229, 273, 285
DTMRun, 381–383
dual, 328

East, 242
easy, 68
edges, 228
else, 109
ends, 219
entries, 42, 43
Eulerian, 269
Euler, 269, 270
evalb(p(3)), 34
evalb, 34, 156, 166, 326
evalf, 132
eval, 70, 288
evalm, 85
evals, 153
eweight, 313
expand, 121
ExtendTree, 315

factorset, 102
FAIL, 327
Family, 276
Fibonacci2, 154
Fibonacci, 154
fibonacci, 20, 173, 174
FindNumOfOnes, 384
FindPath, 254
finished, 129
firstpart, 131
floor, 49
flow, 263, 264
foo, 9
forget, 148
for, 10, 12, 13

G2, 231
gcd, 64, 95
GenerateKeys, 81, 83, 96
GreedyColor, 267
gsimp, 221, 253, 275
gunion, 244

Hanoi, 149, 150
Hash, 72
HasseDiagram, 210
help, 9, 341
HuffCompare, 287
Huffman, 316

IC, 114
icosahedron, 236
Id, 47
ifactor, 67, 69, 94, 101
ifactors, 139
if, 10
igcdex, 65
igcd, 63, 64, 95
ilcm, 64
I, 160
implies, 30, 99
incidence, 247
incident, 310
indegree, 230
indices, 42, 43, 362
induce, 240
infix notation, 292
inorder traversal, 290
Install, 72
interface, 70
intersect, 166
Invariants, 249
invphi, 71
irem, 16, 35, 176
IsAntiSymmetric, 194
IsConnected, 275
IsDFArecog, 358
IsDFSMoutput, 356, 358
IsDTM, 381
IsLattice, 206
IsLessThan, 290
IsMeetSemiLattice, 206
Isomorph, 249, 251, 266
isplanar, 261, 268, 269
isprime[i], 175
isprime, 15, 68–70, 174–176
isqrt, 176
IsSimple, 275
IsSymmetric, 194

IsTmTape, 378
IsTree, 275
iterstructs, 118, 129
ithprime, 69, 111, 175

Kruskal, 322

lastpart, 131
lcm, 64
LCPRNG, 76
length, 49, 338
linalg, 20, 58, 85, 173, 190, 192, 214
Linear, 209, 210, 222, 223
ListInd, 308
listlist, 288
ListSum, 308
logic[bsimp], 335
logic, 32, 33, 99, 324, 326, 328, 329,
    335, 341, 342
Lookup, 72, 73, 75

macro, 72
MakeDigraph, 190
MakeMatrix, 201
ManyFunctions, 60
map, 34, 40, 42, 89
MATHISFUN, 78
matrix, 85, 89, 192
MaxAndMin, 58
maxdegree, 267
MaximalElements, 202
mcoeff, 125, 126
Merge, 298
MergeSort, 299
mersenne, 92
mindegree, 267
MinimalElements, 204
minus, 166, 167
MinWeight, 310
mkRelation, 215
MOD2, 330, 342
mod, 94
modp, 41
move, 365, 367, 370
MovePiece, 318, 319
msolve, 63
msqr, 96

multinomial, 123

ndisks, 150
neighbor, 246
networks, 190, 191, 208, 217, 222,
    273, 274
new session, 46
new, 191, 218
next, 129
nextpart, 131
nextprime, 69, 70, 82, 91, 93, 111
nextvalue, 129
nops, 50, 166, 351
norm, 58
NQueens, 306, 307, 318
NULL, 16
numbcomb, 118–120, 136
numbpart, 117, 130, 131
numbperm, 117, 128
numtheory, mersenne, 92
numtheory, 53, 67, 71, 92, 97, 102,
    205
NumToUpper, 77

octahedron, 236
od;, 12
op, 42, 80, 124, 340, 351
or, 27, 324–326, 334
outdegree, 230

partition, 130, 131
partitions, 117
path, 254
permute, 50, 117, 128
petersen, 235
phi, 71
postfix notation, 292
postorder traversal, 291
powcreate, 170
powerset, 39, 265, 340
powseries, 170
prefix notation, 292
preorder traversal, 289
prevpart, 131
prevprime, 69, 70, 111
PrimeAfter, 17
Prim, 311, 322

printed, 15
printf, 149, 341
printlevel, 343, 344
print, 13, 30, 341
PrintMove, 149
Prob3, 269
procedure body, 14
proc, 9, 108, 109, 327
ProductOfSumExpansion, 347
pseudorandom, 75

randbool, 341, 342
randcomb, 118
rand, 82, 94
randmatrix, 88, 214
random, 266, 268, 270
RandPairs, 143
randpart, 117, 130, 131
randperm, 117, 128
randpoly, 37
readlib, 60
read, 17, 18
RecSol2, 154–156, 159
recurrence equation, 8
RefClose, 195, 201
Regex2NFA, 378
relation, 183
Release 3, 21
Release 4, 21
remember, 109, 110
restart, 46, 342
RETURN, 55
RSA, 81
rsolve, 8, 150, 159, 160, 164

semicolon, 12–14
seq, 36, 70, 108, 167
shortpathtree, 258
simplify, 127
solve, 51, 78, 151–153, 159
sort, 37, 124, 284
spantree, 254, 274, 276
stirling1, 128
subsets, 129
subs, 7, 106, 153
SubSum, 308

SumOfProductsExpansion, 336, 337,
339, 340, 347
SymmClose, 201
symmdiff, 39
system, 18

tau, 97
tautology, 100
TestPrime, 16
tetrahedron, 236
TheoremVerify, 279, 280
time, 60, 61, 94, 109, 198
tpsform, 171
trace, 58, 343, 344
TransferDisk, 149
Trav, 291
type, 28, 31, 176, 347

unapply, 153
union, 166
Unique, 293, 295
untrace, 343
UpperToAscii, 73
UpperToNum, 73, 77

ValidQueens, 305
vdegree, 230, 278
verboseproc, 45, 70
version, 273

Warshall, 201
weights, 259
while, 13
with, 19, 20, 32, 39, 117, 174, 217
W, 51
writeto, 341

zip, 27